Languages for Digital Embedded Systems

The Kluwer International Series
in Engineering and Computer Science

Languages for Digital Embedded Systems

Stephen A. Edwards

Advanced Technology Group
Synopsys, Inc.

Kluwer Academic Publishers
Boston / Dordrect / London

Distributors for North, Central and South America:
Kluwer Academic Publishers
101 Philip Drive
Assinippi Park
Norwell, Massachusetts 02061 USA
Telephone (781) 871-6600 / Fax (781) 681-9045
E-Mail <kluwer@wkap.com>

Distributors for all other countries:
Kluwer Academic Publishers Group
Distribution Centre
Post Office Box 322
3300 AH Dordrecht, THE NETHERLANDS
Telephone 31 78 6392 392 / Fax 31 78 6546 474
E-Mail <services@wkap.nl>

 Electronic Services <http://www.wkap.nl>

Library of Congress Cataloging-in-Publication Data

Edwards, Stephen A., 1970-
    Languages for digital embedded systems / Stephen A. Edwards.
       p.cm. – (Kluwer international series in engineering and computer science ; SECS 572)
    Includes bibliographical references and index.
    ISBN 0-7923-7925-X (alk. paper)
      1. Embedded computer systems – Programming. 2. Programming languages
(Electronic computers) I. Title. II. Series.

TK7895.E42 E39 2000
004'.33—dc21

                                                      00-056133

Contents

Figures

Preface

Digital embedded systems are electronic devices that perform computation but do not appear to be computers in the run-Microsoft-Office sense of the word. They are ubiquitous, of growing importance, and devilishly difficult to design because of the panoply of problems they must solve. Consider a cellular phone. It must transmit and receive radio-frequency signals, digitally encode and decode a speech signal, handle hand-offs between cells, monitor the status of the call, and interact with a person through a keypad and display. Designing a system that solves such varied problems calls for tools specific to a problem domain, and in particular different languages.

This book presents and contrasts languages commonly used to describe the subsystems in a cellular phone and similar digital embedded systems. Some of these languages are in common use; others have significant theoretical value. Their scope ranges from hardware modeling to digital signal processing. To limit its scope, this book only addresses those languages that manipulate discrete (digital) values. While the design of real embedded systems often involves solving problems outside the discrete domain, solving discrete problems is often the bulk of the design work.

Why are there so many languages and why is this book not about the one language that will solve all embedded system design problems? Unfortunately, no such language will ever exist because solving the broad range of problems in embedded system design demands a general language, yet many questions about a design can only be answered when the design is described in a specialized language. Each language in this book has successfully traded off just enough generality to become tractable and yet still solve a wide range of problems.

Languages shape designs and design styles shape languages. Most languages in this book were developed to facilitate a particular way of designing systems, and many have since affected the design process. Keep this in mind as you read about, say, the model of computation in SDL or the process and interrupt model in a typical RTOS. Both encourage a particular approach to design.

Many language issues discussed here are practical ones related to how a language is compiled or synthesized. While not part of most formal language definitions, such issues are inescapable when designing a real system where size and performance are as important as function. A danger in discussing such issues is that different compilers often produce functionally equivalent results with different performance characteristics. Some even allow the user to trade size for speed. Nevertheless, most languages have a de facto compilation technique that produces similar results across many compilers.

Senior undergraduates, graduate students, and practicing engineers should all benefit from this book. I assume the reader is familiar with one of the hardware or software languages (e.g., C or Verilog).

Drafts of this book have been used in a one-semester embedded systems class taught by Prof. Brian Evans at the University of Texas, Austin. Each chapter ends with a set of simple exercises similar to those used in the class; most review points from the chapter. The class culminated in a design project centered around one of the languages.

Free tools for many of the languages included here are available on the Internet. The Internet Resources chapter lists those I know about; many more are surely available.

While reading this book cover-to-cover will not turn the reader into an expert designer, my hope is that it will provide a broad awareness of the languages used in embedded system design.

Stephen A. Edwards
San Francisco, California
June 2000

Acknowledgements

This book had its genesis when my editor Jennifer Evans contacted my former advisor, Edward Lee, at Berkeley and asked him to suggest authors. Edward evidently thought writing a thesis just wasn't enough (it wasn't: I was still standing at the end), so he suggested Jennifer contact me. If this exchange hadn't happened, you would not be holding this book in your hands.

The next hurdle was the approval of my boss, Don MacMillen, since writing a book like this was not something I would be doing in my spare time. Fortunately for me (and, I hope, for you the reader), he was enthusiastic about the idea and certainly deserves a paragraph in the acknowledgements.

Both the University of Texas at Austin and I have a prize in Prof. Brian Evans. His enthusiasm in the project and willingness to subject his poor EE382C-9 students to drafts of this book certainly made this a better document. Perhaps this will make up for the snide remarks I made about him in the acknowledgements of my thesis.

Fortunately for me, Synopsys is populated with a cast of thousands who also happen to know something about the languages in this book. Randy Harr provided excellent feedback on the VHDL chapter, suggesting the chip, socket, and parts list analogy for entities, architectures, and components. Karen Pieper went over the lengthy Verilog chapter with a fine-toothed comb and helped me pick out the lice. Stan Liao similary picked apart the SystemC chapter. Valeria Bertacco scanned the dataflow chapters faster than I thought possible. Joe Buck explained dataflow scheduling and System Studio to me; Logie Ramachandran also improved the System Studio chapter. Lane Dailey provided additional feedback. Abhijit Ghosh did help me understand SystemC, but mostly he insisted I include him here. Although not part of Synopsys, Rod McLaughlin read the Java chapter for me and pointed out one deprecated language feature after another.

No list of acknowledgements would be complete without a mention of the author's wife, and this is no exception. Not only did Nina provide the standard encouragement, toleration, and other things frequently found in these contexts, she also drew the cover and the marionette you have already started to look at on the next page.

1

Language Basics

Using a language is like operating a marionette. You may not touch the puppet; instead you make it move by manipulating a handle connected to the puppet through strings. Just as a good puppeteer understands how the handle moves the puppet through the strings, a good designer understands how the syntax (the handle) of a language controls the model of computation (the puppet) through the semantics of the language (the strings).

The syntax of a language is what we generally think of first when we think of a language. The syntax is a grammar for the language, defining what "words" the language contains and how they may be combined to form sentences and paragraphs. Such words include keywords such as `begin` and `if`, user-defined identifiers such as `alu`, punctuation `{}`, and numbers. Many of the languages described in

Figure 1.1: Using a language is like operating a marionette. To make the puppet move, you manipulate a handle (use the syntax of the language) to pull strings attached to the puppet (the semantics of the language).

this book are textual, meaning their syntax is defined in terms of strings of characters, but others also have graphical representations.

A language's model of computation defines what things the language can describe and how they behave. For example, the model of computation for a hardware language might consist of a set of transistors and wires that connect their terminals. Once set in motion, the transistors operate concurrently and communicate through wires that transport voltages and currents between the transistors. This differs dramatically from the model of computation of a software language, which might consist of a sequence of statements, a memory, and a program counter. The statement at the program counter reads and writes memory before sending the program counter to the next statement.

The semantics of a language defines how the syntax defines objects within the model of computation. The semantics of a transistor definition might mean to add a transistor and connect its pins. An if statement in software might mean to evaluate an expression and execute its then or else block depending on the result.

One of the goals of this book is to introduce the syntax, semantics, and models of computation of a wide variety of successful and novel design languages. Embedded systems often contain a wide variety of problems and a designer who attempts to solve them all with the same language, and hence way of thinking, is usually doomed to spend too much time ultimately to get it wrong. Choosing the best language for the job requires knowing what languages are available.

The complexity of the languages in this book varies considerably. The KISS format is the simplest: it describes a single finite-state machine as a table of transitions. Its model of computation consists of a state and a collection of transitions that advance the state in response to input patterns. The SPICE format is radically different but nearly as simple: the model of computation is a set of primitive elements whose pins are connected by nets. At the other extreme is the assembly language for the TMS320C6701 VLIW DSP (Section 6.4). Its model consists of registers, memory, eight functional units, a fourteen-cycle-deep pipeline, and an enormous number of resource constraints including inter-register-bank communication paths, memory accesses, and what instructions can be in the pipeline simultaneously.

While the syntax of a language is usually definitive, many languages have been interpreted in more than one way. VHDL and Verilog are typical: they started life as languages designed for input to simulators, but have since been pressed into service as specifications of circuits to feed to logic synthesis. This is a big jump for their models of computation: both were designed on a discrete-event model, but logic synthesis uses a synchronous model. Not every discrete-event construct has an obvious implementation in synchronous circuitry, so the languages have synthesis subsets. The SystemC language reinterprets C++ as hardware in a similar way.

Although this book does not address formal language syntax or semantics, a vast body of literature does. Parsing is the interpretation of a program written as a sequence of characters as words and phrases; the earlier chapters of Aho, Sethi, and Ullman's "Dragon Book" [1] discuss this in detail. A grammar defines a language's syntax by defining rules for interpreting tokens as phrases, e.g., a function definition is a name followed by a list of arguments and a body. A body is a list of statements. A statement is an if statement, an assignment, etc.

In the programming language community, semantics and models of computation are closely intertwined. There have been three main approaches to semantics: operational, denotational, and axiomatic. Operational semantics discuss the actions a running program takes, denotational semantics address the function a program computes, and axiomatic semantics are geared toward proving things about a program's behavior. The field uses very sophisticated discrete mathematics and logic and can be intimidating to the uninitiated. Winskel's book [82] is a good introduction; Gunter [32] is more sophisticated.

A language user generally does not need to understand the formal semantics of the language; an informal presentation (e.g., like those in this book) is usually enough. The same does not apply to a language designer: many of the languages described here were not designed with a formal semantics in mind and are sloppy as a result. In most cases the language designer did not fully appreciate the implications of a particular language construct and was satisfied with whatever behavior it produced in the particular implementation he was working on. Sadly, such examples abound and the language user must be aware of these problems to avoid them.

1.1   Specification versus Modeling

Design languages have traditionally had two fairly different objectives. A language designed for modeling is good at answering "what will it do?", whereas a specification language is good for saying "this is what I want." A model might specify how quickly a logic gate will compute its result, but a specification would not since such delays generally cannot be controlled. Instead a specification might define the maximum delay between an input and an output.

Most hardware languages have been aimed at modeling; specification has been for software. History explains the difference: hardware languages evolved from a need to simulate the behavior of small analog integrated circuits such as operational amplifiers before they were built. Only later did integrated circuits become complex enough to demand specification languages. Models in VHDL and Verilog, originally designed for simulation, are now interpreted as specifications.

Software languages, by contrast, evolved from a need to write machine language faster. Software is much more predictable than a circuit; it is much easier to write a program with a given behavior than it is to coerce a collection of transistors into having that behavior. Instead, managing detail is the main challenge in software. Performing a high-level action requires many small instructions to occur in order.

Time has blurred the differences between hardware and software design. Integrated circuits (ICs) must be built hierarchically to manage detail and the time response of software has become critical in many systems. Now, few interesting systems are exclusively software or hardware; most are built with one or more processors controlling off-the-shelf or custom hardware.

Modeling languages are often declarative; specification languages imperative: a model says "here is what I have," a specification says "here is how I want it done." The difference stems from how they are usually used. When using a modeling language, a designer will often have an idea of what will be built, but does not know how it will behave. A modeling language provides a convenient means for specifying the characteristics of the system so that a simulator can calculate emergent properties such as amplifier gain. By contrast, the user of a specification language wants to describe a complicated system succinctly and let an automatic tool infer the details.

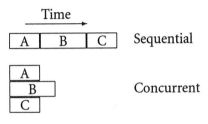

 Sequential

Concurrent

Figure 1.2: Sequential and concurrent behavior. Three actions, A, B, and C, happen one after the other in a sequential system and at the same time in a concurrent system.

1.2   Concurrency and Sequentiality

The contrast between concurrent and sequential behavior appears throughout this book (Figure 1.2). In a concurrent specification, actions may occur at the same time, which has significant implications when they wish to communicate. Solutions usually include a synchronization mechanism that may be local or global.

A sequential specification, by contrast, lists a sequence of actions and the order in which to do them. Generally, the previous action must finish before the next one starts. For we as humans, conceiving and designing sequential behavior is far easier. If action a wants to communicate to action b, simply run a before b. Unfortunately, doing things in order is slower than doing them simultaneously.

Another disadvantage of sequential behavior is that it does not match many processes in the physical world. Electrical systems are naturally concurrent; overhead is required to coerce them into behaving sequentially. This usually takes the form of memory elements whose values change in lockstep with a global synchronizing clock signal. Using a sequential process to control a concurrent one is also a challenge. Assigning a concurrent system process to each concurrent object being controlled is a natural approach, but sequential systems must resort to switching their attention between objects, raising the possibility of missing or at least responding slowly to an event.

Most systems and languages have both sequential and concurrent behaviors. A common style consists of sequential blocks running concurrently. This provides the best of both worlds: the details are described using the easier-to-comprehend sequential style, but the overall system is concurrent, gaining the speed advantage.

1.3   Nondeterminism

A system is deterministic if it will respond in exactly one way to any input. A system that adds two numbers is deterministic. A nondeterministic system can have two or more possible behaviors. This does not mean the system behaves randomly, it just means that it is impossible to predict what it will do.

For example, here are three simple systems

$$d(a, b) = a + b \qquad\qquad \text{Deterministic}$$

$$r(a, b) = \begin{cases} a + b & \text{with probability } 0.25 \\ a - b & \text{with probability } 0.6 \\ ab & \text{with probability } 0.15 \end{cases} \qquad \text{Random}$$

$$n(a, b) \in \{a - b, ab, a + b\} \qquad\qquad \text{Nondeterministic}$$

The d system is deterministic: there is exactly one answer for each pair of inputs. The r system is random: there are three possibilities for any pair of inputs, and we are told how frequently to expect each of them. The n system is nondeterministic. For any pair of inputs there are three possibilities, but we know nothing about how the system will choose among them. System n may behave exactly like system d or r, and simulating n might not help us distinguish it from other systems.

Most of the languages in this book have some form of nondeterminism. In C, for instance, the order in which function arguments are evaluated is implementation dependent, meaning a compiler is free to evaluate them in any order it wants. They are generally not evaluated in a random order. On the contrary: most compilers have a fixed policy that evaluates every function's arguments in the same order. So programs do not behave randomly: a particular compiled version of a program will always produce the same output for the same input. But the output might differ when using a different compiler or the same compiler in a different mode.

Sequential systems are usually deterministic. If each action is well-defined and the sequence in which they are to run is also well-defined, the behavior of the system as a whole is unique. There are no choices about how the system will behave.

Concurrent systems, on the other hand, often have nondeterministic behavior that depends on their speed. Consider two concurrently-running operations attempting to change the same memory location. This is a race, and the slowest operation will set the value last. There is a potential for interference any time two concurrently-running operations try to observe or modify the same resource.

The languages in this book address nondeterministic concurrency in two ways. Some allow nondeterminism but provide mechanisms to avoid it. A Java process can lock an object to gain exclusive access, forcing other threads attempting access to wait until the locking thread has finished with the object. The burden of locking every potentially shared object, however, falls on the programmer.

The other approach is to restrict the language to ensure determinism. For example, Kahn's systems require shared resources to order the processes that wish to access them. Kahn's formalism makes describing certain behaviors difficult, but when the formalism fits, it frees a designer from having to design away nondeterminism.

Nondeterminism can be helpful. Systems described nondeterministically are often easier to simulate and synthesize, and can be more efficient simply because they do not spend time avoiding nondeterminism. Nondeterminism arising from concurrency is natural and arises from the difficulty of controlling the speed of operations.

Checking whether a nondeterministic system behaves correctly is more difficult than the equivalent test for a nondeterministic system, since many more behaviors are possible. Consider five things racing to change the same value. Each of the five could be the fastest, then there are four possibilities for the second fastest, etc. Even for one event, there are exponentially many possibilities; testing a large system with many such events is usually impossible.

Usually, a nondeterministic system description is simulated using a deterministic simulator that consistently simulates one of the many possible behaviors. The system will appear to work, but an implementation that chooses another of the possible behaviors will fail. Testing a system one a variety of simulators (or equivalently, compiling it with multiple compilers) can raise the level of confidence, but this is not a complete solution.

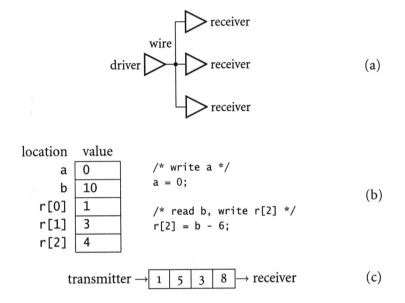

(a)

(b)

(c)

Figure 1.3: The three types of communication used by the languages in this book. (a) Communication over a wire: the driver always sets the value of the wire; the receivers always observe this wire. (b) A memory in a state and statements that read and write to it. (c) A FIFO used for communication. The transmitter sent an 8 followed by a 3, a 5, and most recently a 1. The receiver will read an 8 next, then 3, 5, and 1.

1.4   Communication

The languages in this book use the three broad classes of communication shown in Figure 1.3. Hardware languages use the wire: an object whose value is constantly set by a single transmitter and can be read by any number of receivers. The value changes only when the transmitter decides to send a different value. Transmission through a wire is often modeled as being instantaneous, but models that attempt to determine system timing may include a delay. Each wire's transmitters and receivers are fixed while the system is running, although objects often choose when they transmit and receive. The value on a wire with multiple drivers comes from a resolution function that usually returns the value all drivers agree on, if one exists, and an undefined value otherwise.

Software communicates through memory: a sequence of undifferentiated storage locations that can be read or written. A process may put a new value in a memory location, overwriting an old value, or read the most recently-written value. Unlike wires, memory is a sequential mechanism whose value is observed only when requested.

The dataflow languages use first-in first-out (FIFO) buffers for communication. These are like pipelines: a single driver sends data tokens, and a single receiver reads them out in the order they were written. Some FIFOs limit the number of unconsumed tokens in the buffer, others allow them to grow without bound. A read may fail when there are no new tokens to consume, and a write may fail if the buffer is full.

1.5   Hierarchy

Most languages in this book provide a way to define systems hierarchically, that is, to define new objects by combining other objects (Figure 1.4). This simple concept is one of the most powerful techniques for designing large, complex systems.

A hierarchical object often has specific inputs and outputs, which helps to make systems described hierarchically easier to build correctly, test, and modify because parents only have limited visibility into their children. Such barriers partition the design into regions that can be built, tested, and modified separately. A narrow interface between child and parent is easier to understand.

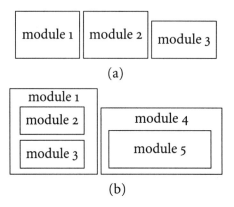

(a)

(b)

Figure 1.4: Systems (a) without and (b) with hierarchy.

It is difficult to underestimate the utility of an interface-based approach to designing systems. Consider the worldwide telephone system, perhaps the most complex system on the planet. The two-wire telephone interface in our homes has remained unchanged for nearly a century, but in that time, the electronics on either end of the line has completely changed a few times. The system continues to work, and no change at the phone company has ever forced everyone to buy new phones. Telephone (cordless phones, computer modems) and switching technology (digital lines, switches, optical fibers) have evolved simultaneously and independently.

The other advantage of hierarchy is how it facilitates reusing components within a design. Often, the same problem appears in many places within a system. A hierarchical system need only contain one solution to the problem that is reused as needed. If an error is discovered in the solution, it only needs to be fixed in one place.

Hierarchical reuse is common in software, where it takes the form of a function. A typical function might send a string to an output device. Reuse is common in hardware at the lowest level. Most hardware designs use a library of simple gates, counters, and registers. But more complicated things, such as controllers built as finite-state machines, are usually so problem-specific that they cannot easily be reused.

1.6   The Languages in this Book

This book discusses four groups of languages: hardware, software, dataflow, and hybrids of the other three. Although all languages ultimately describe some sort of computation, the styles they use can differ dramatically. Hardware languages have traditionally been used for modeling systems that will be built, and so have concentrated on describing structure. Software languages have aimed to allow sequences of processor instructions to be described more abstractly. Dataflow languages are alternative way to specify concurrent software systems using communicating processes.

Having the most users, software languages have experienced the most change, reflected in their wide range of abstraction levels. The assembly languages are the most basic, merely providing a symbolic way of listing instructions in the processor's machine language. A machine instruction written 10ef2c38 in hexadecimal might be written add r1, r5, r8 in assembly language. The C language, the most successful software language, provides essentially a structured way of writing assembly language by allowing expressions (a=b+c*d) and control-flow instructions (if (a==2) b=8) to be grouped into sequences of instructions called functions that can be called from multiple places within a program. The C++ language extends C with additional mechanisms for structuring big programs, allowing new object types to be created and restricting the functions that may manipulate those objects. The Java language, based on C++, further shields the programmer from seeing (and hence disturbing) the details of the processor's operation by prohibiting direct pointer manipulation and providing automatic garbage collection of disused memory.

Operating systems are included with the software chapters because they can add concurrency to traditional software languages. An operating system's scheduling policy can greatly affect system behavior, so special attention is paid to the difference between priority-based scheduling designed to meet deadlines in real-time operating systems (RTOSs) and fair scheduling in timesharing systems.

The hardware languages have evolved by borrowing many ideas from the software community. The SPICE format is like an assembly language: it is little more than a list of components and how they connect, but it does provide hierarchical specification. The other two hardware languages, Verilog and VHDL, started life with very different objectives but have since been pushed into nearly identical roles as input for logic synthesis. Verilog's goal was to replace gate-level simulators with a unified language able to describe gates, testbenches, and abstract behavior. VHDL was intended as a language for modeling a wide variety of digital systems ranging from the gate to system level. Both run concurrent procedures (like C functions) in a SPICE-like structural hierarchy.

Specifying signal processing systems is the main use of dataflow languages, which describe concurrent software with processes that communicate through FIFO buffers. Kahn's process networks are general and deterministic, but such generality comes at the cost of significant run-time scheduling overhead. Fortunately, Kahn's networks can be restricted to the Synchronous Dataflow subset, whose restricted communication patterns allow compile-time scheduling to produce very efficient implementations, and is general enough to specify most signal processing algorithms.

The languages in the hybrid part of this book mix features from the other languages. Esterel combines the concurrent, synchronous clock-based semantics of digital hardware with the high-level control constructs of a language like Java. SystemC is a style of C++ designed for simulating and synthesizing synchronous digital hardware. The Polis model aims for efficient implementation in hardware and software. Its systems contain finite-state machines communicating through single-place buffers. An SDL system also consists of communicating finite-state machines, but SDLs use buffered communication. CoCentric System Studio combines finite-state machines, dataflow, and Esterel-like synchronous communication.

A language is successful for many reasons, but one of the more important is that it allows designers to craft good solutions. Compilation and synthesis have not evolved (and, for theoretical reasons, never will) to where every system description that produces the same results will be implemented optimally. No compiler is always able to recognize an inefficient bubble sort algorithm and replace it with the much more efficient quicksort. Such optimizations are and will remain a human activity, so successful languages must provide a strong, understandable coupling between specification and implementation.

1.7 Choosing a Language

Choosing the right language depends on many factors. In theory, you should understand the problem you wish to solve, consider how its solution would look in each language, and choose the language with the most elegant and efficient solution. In practice, more practical reasons often dominate. Available language tools, your knowledge of a language, and the traditions of an organization are usually deciding factors. Nevertheless, it helps to understand the available languages.

One of the main choices is whether to implement a solution in software, hardware, or a combination. Hardware is generally more costly to design and manufacture in small quantities, but it is the solution of choice where speed or power is critical, or for systems that must interact with the physical world. Software is more flexible, generally easier and cheaper to develop, and its "manufacture" (i.e., copying data) is often free, but it is usually slower and more power-hungry than equivalent hardware. A growing majority of systems use hardware for speed-critical tasks and interfaces and software for the rest.

Whether control or data dominates a problem is another important dimension. The SDF language is excellent for expressing algorithms that perform uniform arithmetic on a continuous stream of data, i.e., signal processing algorithms. At the other extreme, Esterel provides concurrency, preemption, exceptions, and other high-level control constructs that is apt at describing systems that must react to many different simultaneous trigger-like inputs. Other languages fall between these two extremes.

For software the main choice is whether to use assembly or a high-level language. Compilation technology has rendered assembly unnecessary for all but the most speed- or size-critical applications, or those that require direct access to processor features. Unfortunately, compilers are unable to generate efficient code for certain processors, notably some DSPs and small microcontrollers, requiring the use of assembly. Programming in assembly provides precise control over processor behavior, making the very best solutions possible, but such control comes at the price of tedium: programming in assembly is generally slow and error-prone. In situations where assembly is necessary, a good compromise is to use assembly for speed-critical portions and C or C++ for the remainder. Most operating systems are written this way. Among the higher-level languages, C produces the most efficient code and is extremely portable: a C compiler exists for virtually every processor. C++ is better for large projects, but can generate larger executables and is a harder language to use correctly. Java is less error-prone than C or C++, but is less efficient and remains less portable than C or C++.

Verilog and VHDL are the dominant languages for hardware. Verilog is more succinct with more built-in facilities for describing digital logic. VHDL has fewer logic-specific constructs but is more flexible for modeling higher-level behavior. Verilog may be the better choice for logic synthesis, VHDL for broader modeling tasks.

SystemC is emerging as an alternative to Verilog or VHDL for modeling and synthesizing hardware. While somewhat more clumsy, it provides access to all the facilities of C++ and promises better simulation speed. It is a better choice for methodologies that start with a C or C++ model and eventually convert it to hardware.

SDL is good for communication protocols where data arrives in orderly, discrete packets. The Polis model blurs the boundary between hardware and software, helpful when partitioning decisions need to be delayed. Esterel excels at specifying complex, concurrent control behavior. SDF is elegant for many pure signal processing applications. System Studio combines SDF and Esterel semantics to provide a better means to describe systems that demand both a substantial dataflow component and significant control behavior.

1.8   Exercises

1–1.   What is nondeterminism? How might nondeterminism arise? (give two examples) What are the advantages of nondeterminism in a software language? The disadvantages?

1–2.   Why is it important that a language provide precise control over how a system is implemented?

1–3.   In what circumstances would you choose to use an assembly language instead of a high-level software language?

1–4.   When would you choose a hardware implementation? A software implementation? A combination?

Part I

Hardware

A hardware language describes an electrical circuit: a collection of concurrently-operating components communicating through wires. In a digital circuit, these components are gates that compute a Boolean logic function (say, AND) and state-holding flip-flops whose inputs are sampled and held when their clock signal rises. The hierarchically-defined block is the main structuring mechanism: the contents of a block, which includes components and instances (copies) of other blocks, is isolated from the rest of the design. Arrays of wires (busses) and arithmetic expressions can describe regular structures. High-level hardware languages also allow software-like procedural descriptions that are executed in simulation or translated into gates and wires.

2

Hardware Basics

Digital systems represent information using discrete values to make them resistant to noise, manufacturing variation, and aging. Consider the challenge of guessing the position of a light dimmer by observing the light it controls. You would have to ignore other sources of light in the room, consider whether the light has dimmed with age, and guess whether someone had switched your hundred-watt bulb for a seventy-five watt one when you looked away. Now consider how much easier it would be if you knew the light was controlled by a switch that was on or off. This is the advantage of digital systems.

The digital systems discussed in this book are built with integrated circuits containing wires and transistors. The transistors are voltage-controlled switches, and the wires steer voltage between transistors. Values are represented with two voltages. Although voltage is a continuous value, the transistors treat all voltages near "off" as completely off and all voltages near "on" as completely on.

Connected switches can compute simple logic functions. Switches connected in series compute a logical AND function, as shown in Figure 2.1. That is, a connection is made only if the first switch *and* the second are conducting. Microwave ovens use a circuit like this for

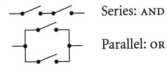

Series: AND

Parallel: OR

Figure 2.1: Series and parallel switches compute the basic logic functions.

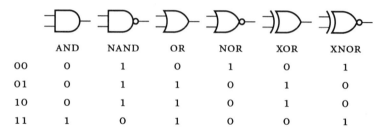

	AND	NAND	OR	NOR	XOR	XNOR
00	0	1	0	1	0	1
01	0	1	1	0	1	0
10	0	1	1	0	1	0
11	1	0	1	0	0	1

Figure 2.2: Symbols, names, and truth tables for two-input logic gates. The bubble on the output of some gates indicates the value of the output is inverted.

safety. The microwave emitter (magnetron) can only operate when the door is closed and the oven is activated.

Figure 2.1 shows parallel switches computing the OR function. A connection is made if the first switch *or* the second is conducting (or both). A phone with on-hook dialing might contain this circuit: the phone is on if the handset is lifted or if on-hook dialing is selected.

Most modern circuits are built using Boolean logic gates, such as those listed in Figure 2.2. A gate is a small group (perhaps four or six) transistors connected so that its output is a logical function of its input (e.g., the AND function). Furthermore, each gate amplifies so long strings of them may be connected without signal degradation.

The NAND (not-and) gate is the building block of modern CMOS integrated circuits because it can be implemented efficiently (just four transistors as shown in Figure 2.5) and can be used to compute any Boolean logic function. The output of a NAND gate is one unless both inputs are one. Shorting both inputs gives an inverter. Inverting a NAND's output gives AND. Inverting all inputs and outputs gives OR.

Multiple-input gates can be built with single input gates. For example, connecting the output of a two-input AND to one of the inputs of another two-input AND gives a three-input AND.

It is easy to construct any logic function using ANDs, ORs, and inverters. Start with the truth table (see Figure 2.3), and consider the input patterns that produce a 1. For each of these, build an AND gate that is true when the pattern appears, then OR together the outputs of all these AND gates to produce the function. This structure is known as a programmable logic array or PLA, and was a typical way to con-

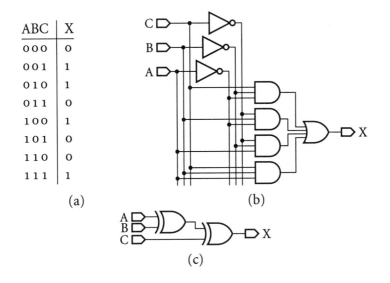

ABC	X
0 0 0	0
0 0 1	1
0 1 0	1
0 1 1	0
1 0 0	1
1 0 1	0
1 1 0	0
1 1 1	1

(a)

(b)

(c)

Figure 2.3: Building a PLA from a truth table. Each true minterm in the truth table (a) (e.g., A=0, B=0, and C=1) becomes an AND gate in the PLA (b). The OR gate combines these to produce the function, but this function has a better implementation (c) in XORs.

struct logic functions before multilevel synthesis and optimization technology was developed.

Digital logic design is a mature field that has many excellent undergraduate texts. I recommend Wakerly [80], Katz [47], and Mano [57]. Weste and Eshraghian [81] is standard for integrated circuit design. Horowitz and Hill's classic [39] is geared toward practical electronic design, including many analog issues.

2.1   Schematic Diagrams

Electrical engineers have drawn schematic diagrams for decades if not centuries. Long a semi-formal means of human communication, many so-called schematic capture programs now allow schematics to be entered on a computer and translated into other formats, including SPICE (Section 2.2), Verilog (Chapter 3), or VHDL (Chapter 4).

A schematic diagram is a drawing that shows how a collection of objects is connected. It consists of symbols and nets connecting pins

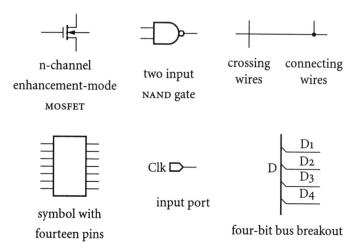

n-channel

enhancement-mode

MOSFET

two input

NAND gate

crossing   connecting

wires   wires

symbol with

fourteen pins

Clk ▷—

input port

D

D1

D2

D3

D4

four-bit bus breakout

Figure 2.4: Typical schematic symbols.

on the symbols. A symbol represents a device such as a transistor, a
NAND gate, or another schematic diagram (see Figure 2.4). Symbols
have pins that are connection points for the nets. Each net connects
two or more pins. A net is something like a physical equals (=) sign.
Electrically, it is a constant potential region, meaning each connection
to the net instantly sees the same value (voltage).

Nets are drawn as straight lines that are usually horizontal or verti-
cal. When two lines touch or cross, they do not connect unless a small
dot (a "solder dot") is drawn at the intersection.

A net is often given a name. Some schematic capture systems consider separated nets with the same name to be connected; others consider this an error. Designers often use the former facility to connect
global nets such as power, ground, or clocks.

A port is a named symbol that represents a pin on the schematic's
symbol. A net connected to the pin on the symbol also connects to
a net connected to the port. Ports are sometimes given a direction,
either input or an output.

A bus is a shorthand for a set of nets. Ports and pins may be labeled as busses; an individual net may connect to a bus (sometimes
using the connect-by-name facility) allowing individual signals to be
broken out. Such busses usually communicate binary numbers.

2.2   The SPICE Format

The SPICE format is a simple way to represent hierarchical schematics. Originally an input to the SPICE (Simulation Program with Integrated Circuit Emphasis) simulator developed at the University of California, Berkeley beginning in about 1975 [62], it was intended to describe analog integrated circuits such as operational amplifiers.

A SPICE file defines a collection of components and how they connect. Each component is declared on a single line. E.g.,

```
M1  Y  A  Vdd  Vdd  P
```

defines an instance named M1 of a P-type transistor connected to nets Y, A, and Vdd. The first character on the line dictates the component type: M is a MOSFET, V is a voltage source, X is an instance of a subcircuit, etc. The nets connected to the component's pins follow the name, and the line ends with name of the model or subcircuit.

The .SUBCKT directive names a new component and its pins. Its body lists the components within; nets there are either pins or local to the subcircuit.

```
.SUBCKT  NAND  A  B  Y  Vdd  Vss
.ENDS
```

The SPICE file in Figure 2.5 defines a NAND gate consisting of four transistors, connects four instances of this NAND gate, and defines a testbench of a power supply and a pair of out-of-phase square waves with 40 ns periods. The .TRAN and .PLOT lines instruct the simulator to perform transient simulation for 160 ns and plot the results.

Transistors on a chip often have similar electrical characteristics. The .MODEL lines define these characteristics and the transistor instance refer to the models they define. This simple example uses the default values of all parameters but a more detailed simulation would specify, e.g., threshold voltages, on these lines.

The SPICE format was never intended to be a standard, just to solve the problem at the time. Years of extensions have left its syntax with plenty of idiosyncrasies. It was originally used with punch cards, so input files are known as SPICE decks. Each component is on a separate card (line). Lines that begin with + are considered part of the last line, making it easy to add cards without modifying existing ones.

```
An XOR built from four NAND gates

.MODEL P PMOS
.MODEL N NMOS

.SUBCKT NAND A B Y Vdd Vss
M1 Y A Vdd Vdd P
M2 Y B Vdd Vdd P
M3 Y A X    Vss N
M4 X B Vss Vss N
.ENDS
```

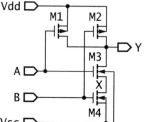

```
X1 A B I1 Vdd 0 NAND
X2 A I1 I2 Vdd 0 NAND
X3 B I1 I3 Vdd 0 NAND
X4 I2 I3 Y Vdd 0 NAND

**** Testbench:

* Power supply
V1 Vdd 0 3.3

* Input waveforms
VA A 0 PULSE(0 3.3 0    0 0 20ns 40ns)
VB B 0 PULSE(0 3.3 10ns 0 0 20ns 40ns)

**** Simulation commands:

.TRAN 1NS 160NS
.PLOT TRAN V(A) V(B)
.PLOT TRAN V(Y)

.END
```

Figure 2.5: A SPICE file describing an exclusive-OR gate built from four NANDs. The schematics on the right illustrate the meaning.

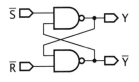

Figure 2.6: An SR latch implemented in NAND gates.

2.3   Sequential Logic

Combinational logic does not remember anything. Its output is always a function purely of it current input. Some systems are purely combinational systems, but most interesting ones have memory.

A latch remembers its history. Figure 2.6 shows the simplest latch, which holds its state when its inputs are both 1, but sets Y to 1 if S goes low and sets Y to 0 if R goes low. Since the output always responds to its inputs instead of a clock signal, this latch is called level-sensitive.

A flip-flop changes state in response to a clock edge rather than a level on its inputs. The D flip-flop in Figure 2.7 is the simplest. The Q output holds its value until the Clk input rises. The value on D when the clock rises becomes the new value of Q until the next clock.

Figure 2.8 illustrates the behavior of the six-NAND flip-flop. When the clock is low, the output latch stores the previous value. When the clock rises, the output latch is set to the previous value of D, but once it has risen, further changes on D are not propagated to the output.

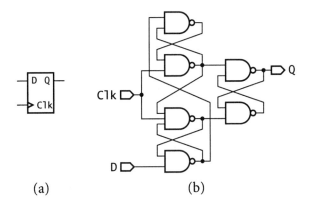

(a)   (b)

Figure 2.7: (a) A positive-edge triggered D flip-flop. (b) An implementation using NAND gates.

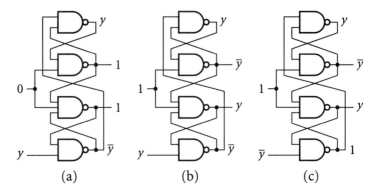

Figure 2.8: The behavior of the six-NAND D flip-flop. (a) When the clock is low, the output latch holds its value. (b) Raising the clock sets the output latch to y: the value of the D input. (c) Once the clock is high, changing D to $\bar{y}$ does not change the controls to the output latch.

This circuit has definite setup and hold times. The D input needs to be stable slightly before to slightly after the clock rises to ensure its value is copied correctly. If D changes during this time, it is not obvious whether the old or new value will be latched. These times are a complex function of the delay of the gates.

The complex behavior of this very common circuit (a big microprocessor might contain ten thousand) is motivation for modeling circuits abstractly. A simulator can more efficiently model this circuit by copying the value of D to Q when the clock rises. This also allows setup and hold times to be checked explicitly.

2.4   Finite-state Machines: The Traffic Light Controller

D flip-flops are used to implement finite-state machines as in Figure 2.9. A finite-state machine has inputs, outputs, and a bank of flip-flops that hold its state. A block of combinational logic computes the outputs and next state as a function of the inputs and the current state. A single clock signal is connected to all of the flip-flops. When the clock rises, the machine advances to the next state.

Finite-state machines are often drawn using bubble-and-arc diagrams such the traffic-light controller in Figure 2.11. It controls a traffic light at the intersection of a busy highway and a farm road (Fig-

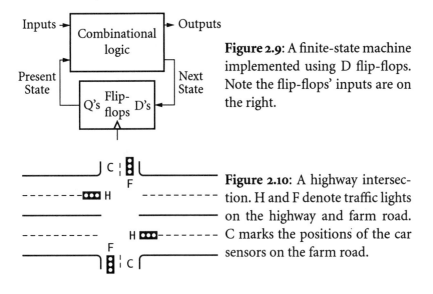

Figure 2.9: A finite-state machine implemented using D flip-flops. Note the flip-flops' inputs are on the right.

Figure 2.10: A highway intersection. H and F denote traffic lights on the highway and farm road. C marks the positions of the car sensors on the farm road.

ure 2.10). Normally, the highway light is green but if a sensor detects a car on the farm road, the highway light turns yellow then red. The farm road light then turns green until there are no cars or after a long timeout. Then, the farm road light turns yellow then red, and the highway light returns to green.

This machine has four states, each corresponding to a particular state of the two lights. These are the circles labeled HG, HY, etc.

The inputs to the machine are the car sensor C, a short timeout signal S, and a long timeout signal L. The outputs are a timer start signal S, and the colors of the highway and farm road lights.

The notation on each transition indicates the conditions on the

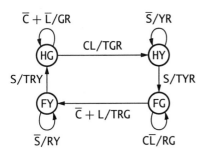

Figure 2.11: A bubble-and-arc diagram for the traffic light controller. Transitions are labeled "input condition/output values". C is the car sensor, S and L are timeouts, T starts the timer, R, Y, and G indicate colors. (after Mead & Conway [59, p. 85])

inputs that fires the transition followed by a slash and the outputs generated by firing the transition. An expression like $\overline{C} + \overline{L}$ means the transition fires if no car is sensed or if the long timeout has not occurred. The TGR output means to start the timer and make the highway light green and the farm road light red.

The controller is connected to a timer. The S output starts the timer, which first produces a short timeout used to time the yellow lights and later a long timeout used to limit the length of the farm road light and guarantee the length of the highway light.

The traffic-light controller is a Mealy machine because its outputs depend both on its inputs and its state. The outputs of a Moore machine, by contrast, only depend on its current state. Mealy machines have the advantage of reacting faster to their inputs (i.e., in the same cycle as the input arrives), but Moore machines are simpler. An advantage of Moore machines is their ability to interact safely. Because a Mealy machine can respond immediately to its inputs, two Mealy machines can communicate back and forth within the same clock cycle. This can lead to unexpected behavior that can depend both on the machine's function and how it is implemented in hardware.

Most logic synthesis systems restrict FSMs to Moore machines to ensure proper timing. Specifically, only state bits may be outputs. Composing such machines is easier because each machine's inputs arrive just after the clock because they are sent directly from other machines. This gives the next-state logic nearly a full clock cycle to compute its result, regardless of the speed of other machines. This is not as straightforward when interconnect delays dominate.

Finite-state machine theory is discussed in many places. Any introductory digital design textbook [80, 47, 57] will have sections on finite-state machines, as will advanced books on logic synthesis [33, 22, 24]. Hopcroft and Ullman [38] is the standard reference on finite-state machines and their ilk in the computer science domain. Finally, a pair of books by Kam, Villa, et al. discusses finite-state machine synthesis at length. The first volume [45] discusses issues such as state minimization between machines; the second volume [79] discusses encoding and how to synthesize the logic for these machines.

2.5   The KISS Format

The KISS format is a simple machine-readable way to represent a finite-state machine destined for hardware. Originally for FSMs implemented in programmable logic arrays (PLAs), it is the simplest language in this book. It describes a single finite-state machine, explicitly listing states and transitions. Alone, the KISS format is not enough to describe large systems since it cannot connect FSMs.

A KISS file (Figure 2.12) is a list of transitions. Each transition is an input pattern, a present state, a next state, and an output vector. Each cycle, the machine is presented with a vector of 1's and 0's that selects a transition from the current state matching the input pattern. The transition sets the outputs and advances the state of the machine.

Input patterns in the KISS format may contain unspecified bits to compactly represent many different values. Each input pattern is a string of 0's, 1's, or -'s, one per input. A - ignores the value of its input, i.e., the input could be 0 or 1 and still match. For example,

```
0--   HG   HG   00010
```

triggers if the machine is in state HG and the farm road car input is 0, regardless of the other two inputs. A - in hardware is trivial—the corresponding input simply is not checked.

The KISS file in Figure 2.12 contains the traffic-light controller from Figure 2.11. The light color outputs have been encoded as a pair of bits. Note how some transitions in Figure 2.11 need two lines in the KISS file because they represent OR conditions.

Implementing the FSM of a KISS file in hardware requires encoding the symbolic states into zeros and ones. Giovanni De Micheli's KISS program [23], which originated the KISS format, did this encoding to minimize the size of the FSM implemented in a PLA (see Figure 2.3). Multilevel logic has since supplanted two-level logic.

```
# Highway and Farm Road Traffic Light Controller

# States: HG, HY, FG, FY
# Inputs: car sensor, long timeout, short timeout
# Outputs: start timer, highway light, farm light
# Encoding: 00 green, 01 yellow, 10 red

#CLS  ps  ns   THiFa

0--   HG  HG   00010
-0-   HG  HG   00010
11-   HG  HY   10010
--0   HY  HY   00110
--1   HY  FG   10110
10-   FG  FG   01000
0--   FG  FY   11000
-1-   FG  FY   11000
--0   FY  FY   01001
--1   FY  HG   11001
```

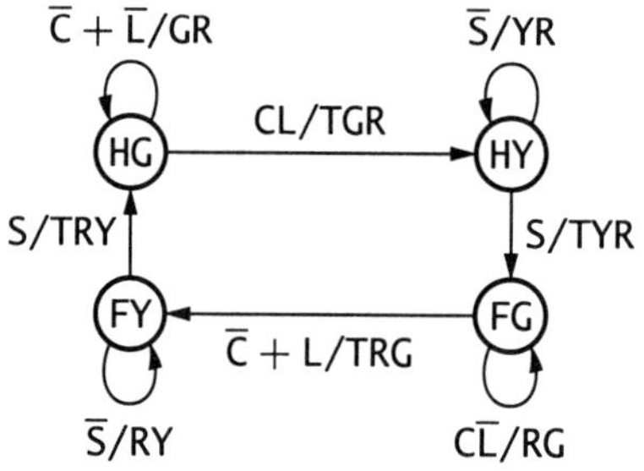

Figure 2.12: A KISS file describing the traffic-light controller. (The bubble-and-arc diagram is a reproduction of Figure 2.11.)

2.6   Exercises

2–1.   What is the main difference between Mealy and Moore finite-state machines? What are the advantages and disadvantages of each?

2–2.   How many rows are in a truth table of n inputs?

2–3.   How many Boolean functions are there of 4 inputs? Of n inputs?

2–4.   (a) Draw a bubble-and-arc diagram for the user interface of a CD player. Inputs: power on, power off, eject, stop, play, and pause. Outputs: eject tray, spin disc, play music. (b) Write this in KISS format.

2–5.   A seven-segment decoder converts the first ten four-bit binary numbers into seven on/off signals controlling the seven segments of a numeric display. (a) Write a truth table for the "b" segment. (b) Draw a PLA that implements this function. (c) Simplify the logic: reduce the number of gates. (d) The output for inputs above ten is irrelevant. Can you use these "don't-cares" to further simplify the logic?

0 1 2 3 4 5 6 7 8 9

```
   a
 f | g | b
 e |_d_| c
```

2–6.   (a) Write a SPICE file describing the positive-edge triggered flip-flop in Figure 2.7b. (b) Simulate it in SPICE and show the output only changes when the Clk input rises.

3

Verilog

Verilog began life in 1984 as an input language for an event-driven simulator and has since become a key language in chip design because it is succinct, expressive, simulates quickly, and much of it can be automatically converted to hardware. It allows systems to be described using structural elements to closely match an implementation or behaviorally for additional simulation speed and easier specification.

Verilog's structural subset resembles many netlist formats: a system is a set of modules that each have ports and contain instances of gates, transistors, or other modules that communicate through wires (nets) within the module. The behavioral subset adds initial and always blocks: concurrently-running sequential processes described using C-like syntax and semantics that write to memory (registers).

Verilog's different modeling mechanisms tradeoff simulation accuracy for speed. Switch-level structural modeling is the most detailed, but because its primitives are bidirectional, its simulation must solve equations. At the other extreme, a behavioral model is written and simulated (executed) like a traditional software program. Behavioral instructions are executed in sequence and registers, where instructions write their results, are simply memory locations, and do not exactly correspond to latches or flip-flops. Furthermore, concurrently-running behavioral blocks may run in any order. While this speeds simulation, it permits nondeterminism.

Generating hardware from structural Verilog is straightforward (it is a hierarchical netlist); generating hardware from behavioral descriptions is harder and is done in two ways. Register-transfer level (RTL) synthesis is the more common: it ignores delays, instead interpreting the behavior of the system between clock cycles. RTL prescribes the number of registers and how and in which clock cycles data moves between them. By contrast, behavioral synthesis determines the number of registers and operators from constraints by performing a scheduling and allocation step before generating hardware. (DeMicheli's book [22] provides an introduction to the algorithms involved.) For example, in RTL, an expression like $e = a * b + c * d$ means the register e will be loaded with the output of an adder fed by two multipliers driven by registers a, b, c, and d. Behavioral synthesis may do this in one, two, or three cycles, use one or two multipliers, and may use an additional register to store an intermediate result.

The standard document, IEEE 1364-1995 [42] defines the Verilog language and, since it was based on a user's manual, is fairly readable, but there are better starting points. Thomas & Moorby's book [76] has steadily improved since its first edition. Moorby designed much of the language. The standard includes the programming language interface, also described by Mittra [60] and Sutherland [74].

3.1   Two Examples

Figure 3.1 is a Verilog version of the XOR built from four NAND gates (Figure 2.5) using a combination switch, gate, and behavioral modeling. The NAND module uses switch-level transistor primitives and the XOR module instantiates NAND four times to build an XOR gate. The internal wires i1, i2, and i3 are implicit in the XOR module.

The testbench module shows how behavioral code can be used to specify testbenches. The module declares registers used by the behavioral always block as stimulus to both the device under test and a reference implementation done as a continuous assignment.

Assign y1=a^b is a continuous assignment, meaning y1 is set to the XOR of a and b. The expression is reevaluated if either change.

The initial block, executed once at the beginning of the simulation, starts a monitor that prints signal values when any change.

The always block generates a waveform by assigning to registers a and b. Each #10 suspends the block for ten time units.

```
module NAND(y, a, b);
  output  y;
  input   a, b;
  supply1 vdd;
  supply0 vss;

  pmos M1(y, vdd, a), M2(y, vdd, b);
  nmos M3(y, x, a),   M4(x, vss, b);
endmodule

module XOR(y, a, b);
  output y;
  input  a, b;

  NAND X1(i1, a, b),
       X2(i2, a, i1),
       X3(i3, b, i1),
       X4(y, i2, i3);
endmodule

module testbench;
  reg a, b;
  wire y1;

  XOR x1(y2, a, b);    /* Device under test */

  assign y1 = a ^ b;  // Reference device

  initial
    $monitor($time,,"a: %d b: %d y1: %d y2: %d", a, b, y1, y2);

  always
    begin
      #10 a = 0; b = 0;
      #10         b = 1;
      #10 a = 1;
      #10         b = 0;
    end
endmodule
```

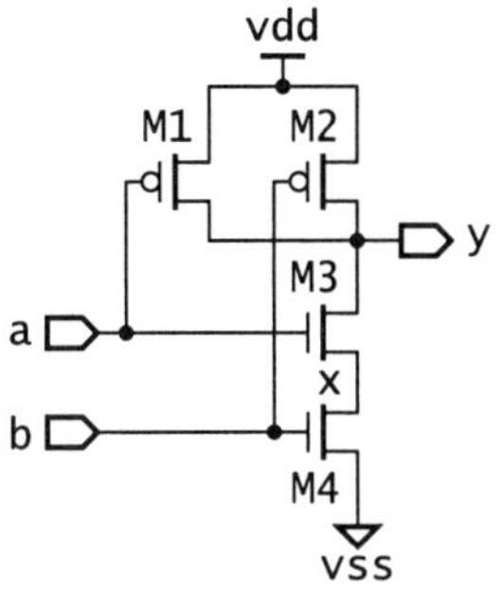

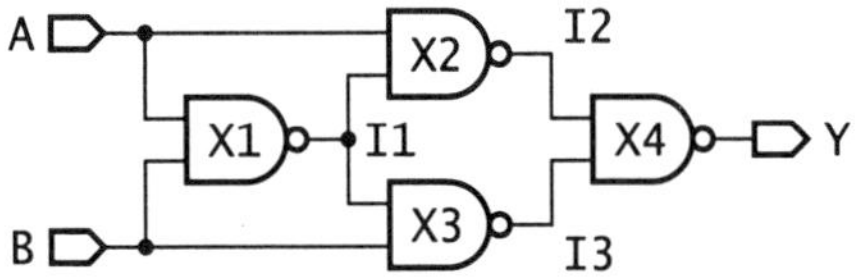

Figure 3.1: A Verilog model for the XOR built from four NANDs. From top to bottom, the three modules use switch, gate, and behavioral modeling. Compare this to Figure 2.5.

```
module top;
  wire       long, short, start;
  wire [2:0] highway, farm;
  reg        clk, car;

  timer t(long, short, start, clk);
  fsm   f(start, highway, farm, car, long, short, clk);

  initial
    begin
      clk = 0;
      car = 0;
      $monitor($time,,"h: %d f: %d c: %d", highway, farm, car);
    end

  always #10 clk = ~clk;
  always #50 car = ~car;
endmodule

module timer(long, short, start, clk);
  output long, short;
  input  start, clk;
  parameter shortTime = 3, longTime = 10, width = 4;
  reg [width-1:0] count;

  assign long = count >= longTime;
  assign short = count >= shortTime;

  initial count = 0;

  always @(posedge clk)
    begin
      if ( start )
        count = 0;
      else if ( count != longTime ) // avoid overflow
        count = count + 1;
    end

endmodule
```

Figure 3.2: The traffic light controller in behavioral Verilog. (1/2)

```
module fsm(start, high, farm, car, long, short, clk );
   output        start;        // Start timer
   output [2:0] high, farm;   // Light states
   input         car,          // Car at farm road
                 long,         // Long timeout
                 short;        // Yellow timing
   input         clk;
   reg           start;
   reg [2:0]     high, farm;
   reg [1:0]     state;        // Current state

   parameter RED = 3'b001, YELLOW = 3'b010, GREEN = 3'b100;
   parameter HG = 2'b00, HY = 2'b01, FG = 2'b11, FY = 2'b10;

   initial state = HG;

   always @(posedge clk)
     case (state)
       HG:
         begin
           high = GREEN; farm = RED; start = 0;
           if ( car && long )
             begin start = 1; state = HY; end
         end
       HY:
         begin
           high = YELLOW; farm = RED; start = 0;
           if ( short ) begin start = 1; state = FG; end
         end
       FG:
         begin
           high = RED; farm = GREEN; start = 0;
           if ( !car || long )
             begin start = 1; state = FY; end
         end
       FY:
         begin
           high = RED; farm = YELLOW; start = 0;
           if ( short ) begin start = 1; state = HG; end
         end
     endcase
endmodule
```

Figure 3.2 concluded (2/2)

Figure 3.2 shows the traffic-light controller from Section 2.4 written using behavioral constructs. The `top` module instantiates the `timer` and `fsm` modules and generates a periodic clock and car signal for a test bench. The `timer` module uses continuous assignments to generate the `short` and `long` timeouts used by the `fsm` module, and an always block triggered by the clock as a counter.

The `fsm` module uses parameters as symbolic constants to encode both the light color signals and the states. The module is mostly a clock-triggered case statement, a common FSM idiom.

3.2   Modules, Instances, and Structure

A Verilog description is divided into modules, each of which has a name, a set of ports, and a body that may contain wires, instances of gates or other modules, continuous assignment statements, and behavioral initial and always blocks.

Instances of modules or primitives begin with the name of the object being instantiated followed by named instances of that object describing how their ports are connected. Connect-by-order is convenient for objects with a small or variable number of ports, meaning the connection to the first port is listed first, the second next, etc. Verilog also provides named connections, which can be easier to maintain. For example:

```
module instances;
  NAND n1(y1, a1, b1);            // Ordered connection
  NAND n2(.a(a2), .b(b2), .y(y2)); // Named connection
endmodule
```

Modules not instantiated within other modules form the roots of the hierarchy (top-level modules). A single top-level module is common, but multiple top-level modules can be useful for testbenches, assembling designs at the top level, and to separate a module that contains parameter settings (i.e., using the `defparam` statement).

Verilog provides a concise notation for regular arrays of instances. Each arrayed instance has a vector-like name such as `n[2]`. If a vector is connected to a scalar port on the arrayed module, Verilog connects each instance to one of the bits in the vector. For example, if a, b, and y are 2-bit vectors, then these are equivalent:

```
and nd[1:0](y, a, b);
and nd[1](y[1], a[1], b[1]), nd[0](y[0], a[0], b[0]);
```

Instance arrays are a fairly recent addition to the language. VHDL's generate statement, which defines instances algorithmically, is another proposed addition.

If a wider vector is connected to a smaller vector, pieces of the wider vector are connected. This is especially powerful when combined with the vector concatenation operator:

```
module add4(y, cout, a, b, cin);
   output [3:0] y;
   output       cout;
   input  [3:0] a, b;
   input        cin;
   assign {cout, y} = a + b + cin;
endmodule

module add32(y, a, b);
   output [31:0] y;
   input  [31:0] a, b;
   wire   [7:0]  carry;
   add4 ad[7:0](y, carry[7:0], a, b, {carry[6:0], 1'b0});
endmodule
```

When a primitive gate or transistor is instantiated, it may be assigned a rising, falling, and turn-off delay (delay to a high-impedance state), with minimum, typical, and maximum values:

```
buf                b1(a, b);  // Zero delay
buf #3             b2(c, d);  // All delays are 3
buf #(4,5)         b3(e, f);  // Rise=4, fall=5
buf #(5,6,7)       b4(g, h);  // Rise=5, fall=6, off=7
buf #(3:4:5)       b5(i, j);  // Min=3, typ=4, max=5
buf #(3:4:5, 4:5:7) b6(k, l); // rise=3:4:5, fall=4:5:7
```

3.3   Nets, Registers, and Expressions

Four-valued vectors are Verilog's main data type. These vectors are natural for digital hardware: each bit corresponds to a wire, and the four values, 0, 1, x, and z, correspond to binary values, an unknown

wire	tri	Simple connection
wand	triand	Wired-AND function
wor	trior	Wired-OR function
tri0	tri1	Resistive pull to 0 and 1
supply0	supply1	Power supply connections
trireg		Can store charge (for switch-level models)

Figure 3.3: Verilog net types.

or illegal value mostly used during system initialization, and the undriven (high-impedance) state of a tristate bus.

Verilog models hold these four-valued vectors in two types of storage. A net behaves like a physical wire: it either floats or is attached to one or more objects (such as a gate or continuous assignment) that constantly sets its value. A register behaves like a memory location: its value holds until it is set by an assignment in an initial or always block. Arrays of registers model larger pieces of memory. Behavioral code may also use integers, reals, and a time type.

Figure 3.3 lists Verilog's net types. These differ mostly in how their value is resolved when there are multiple drivers. Wire/tri nets (synonyms for the same type, intended as commentary on their use) are the most common: they model a simple net that becomes z when nothing drives it and x when driven by conflicting values. Implicitly-defined nets default to wires. A wand/triand behaves as if all its drivers are inputs to a big AND gate: its value is 0 unless all drivers are 1; a wor/trior net behaves similarly, but with the OR function. The tri0 and tri1 nets model resistive pulldowns and pullups respectively. The supply0 and supply1 model connections to power supplies (i.e., always take that value).

The trireg net models capacitive switch-level behavior. Unlike other nets, a trireg holds its value when it is not driven, and has a strength that represents its capacitance (the available strengths are listed in Figure 3.7). If a switch-level primitive shorts two trireg nets, the stronger net's value takes precedence. A trireg net normally holds its value indefinitely, but if a turn-off delay (the third number) is specified, it defines how long an undriven trireg holds its value before transitioning to x. This models capacitive leakage.

```
wire            a;              // A single wire
tri [15:0]      dbus;           // A sixteen-bit bus:
                                // dbus[15] (msb)...dbus[0] (lsb)
tri #(5,4,8)    b;              // Tristate wire with delay
reg [-1:4]      vec;            // A six-bit register
trireg (small) q;              // A charge storage node with
                                // small strength
integer         imem[0:1023];   // A memory of 1024 integers
reg [31:0]      dcache[0:63];   // A memory of 64
                                // thirty-two bit registers
reg [3:0]       rfile[0:15], b; // A memory of 16 four-bit
                                // registers and
                                // one four-bit register.
```

Figure 3.4: Net and register declarations.

When a net has multiple active drivers, the simulator compares the strongest drives to 0 and 1. If they are equal, the net is assigned the unknown value x. If one wins, the net is assigned 0 or 1. If there are no active drivers, the net is given the high-impedance value z.

Nets and registers are always declared within the scope of a module and may include a width, a delay, a strength, and the size of the array. Figure 3.4 has examples.

Verilog models manipulate the values in nets and registers with expressions. Expressions are built from literals (numbers and strings), operators (such as plus or logical AND), and references to objects (nets, registers). Naturally, Verilog's expressions are geared toward operations that are common or useful in hardware.

Literal numbers may be specified in decimal (511), hexadecimal ('hc080), octal (12'o345), or binary (8'b1010_11x1). They may contain x's, z's, and underscores to divide long numbers (underscores are ignored). They may begin with a width specifier. Negative numbers are translated to two's complement, and strings (e.g., "Hello") are interpreted as zero-padded vectors of eight-bit characters.

Figure 3.5 shows Verilog's operators grouped by precedence. The arithmetic operators (+, -, *, /, %) treat vectors as unsigned numbers (i.e., never negative), but treat integers and real numbers as signed.

There are two types of equality. The logical equality operator (==) behaves like a hardware implementation of a comparison. If either operand contains an x or z, the result of the comparison is x. The

+r	(unary)	-r	negation
!r	logical NOT	~n	bitwise NOT
r * s	multiplication	r / s	division
i % j	remainder		
r + s	addition	r - s	subtraction
i << j	shift left	i >> j	shift right
r < s	less than	r > s	greater than
r <= s	...or equal	r >= s	...or equal
r == s	logical equal	r != s	logical not equal
r === s	case equal	r !== s	case not equal
&i	reduction AND	~&i	reduction NAND
i & j	bitwise AND	i ~& j	bitwise NAND
i ^ j	bitwise XOR	^i	reduction XOR
i ~^ j	bitwise XNOR	~^i	reduction XNOR
i ^~ j	bitwise XNOR	^~i	reduction XNOR
i | j	bitwise OR	i ~| j	bitwise NOR
|i	reduction OR	~|i	reduction NOR
r && s	logical AND		
r || s	logical OR		
r ? s : t	conditional		

Figure 3.5: Verilog's operators, grouped by precedence (top is highest). r and s represent any number, i and j represent registers, nets, and integers (not reals).

"case" equality === (so named because its semantics match those of the case statement) requires its operands to match exactly, including any x's or z's, for the result to be 1. The result is 0 otherwise.

The logical reduction operators return a single bit generated by applying the operator between each bit of its vector-valued operand. For example, the expression &a is one only if every bit of the vector a is 1. Similarly, ^a (reduction XOR) returns 1 if there is an odd number of 1's in a.

Single bits and parts of a vector can be addressed like v[2] and v[7:4]. This notation also selects elements of arrays, i.e., a[31]. Individual bits from a vector-valued memory can be selected by copying the desired location to a register then selecting bits from it; no notation does this directly.

Vector-valued expressions may be concatenated to form wider vectors such as {a, 4{q}, b[3:0]}. The notation 4{q} means four copies of the value of q. This notation can be particularly useful on the left hand side of assignments, allowing a function to return a wide vector that is broken apart.

3.4   Gate and Switch-level Primitives

Verilog provides two groups of built-in primitives (Figure 3.6). Gate-level primitives model standard multi-input Boolean logic gates (e.g., those in Figure 2.2). Switch-level primitives model analog effects with transistors that behave as unidirectional or bidirectional switches.

Both gates and trireg nets (see Section 3.3) may be assigned one of the strengths in Figure 3.7 to resolve cases when multiple drivers attempt to set the value of a net. This facility is part of Verilog's switch-level modeling capability, which is now rarely used because it is both too imprecise and too slow for large systems.

3.5   User-defined Primitives

A Verilog user-defined primitive (UDP) is a gate or sequential element defined as a truth table (e.g., Figure 3.8). Compared to expressions or combinations of existing primitives, a UDP is better for complicated sequential primitives (e.g., with multiple clocks) and provides better control over behavior when inputs are undefined.

Name	Inverting	Type	Pins
and	nand	logical AND	(o, i, ...)
or	nor	logical OR	(o, i, ...)
xor	xnor	logical XOR	(o, i, ...)
buf	not	buffer	(o, i)
bufif0	notif0	tristate buffer	(o, i, $\bar{c}$)
bufif1	notif1	tristate buffer	(o, i, c)

Name	Resistive	Type	Pins
nmos	rnmos	unidirectional pull-down	(o, i, c)
pmos	rpmos	unidirectional pull-up	(o, i, $\bar{c}$)
cmos	rcmos	unidirectional pass gate	(o, i, c, $\bar{c}$)
tranif0	rtranif0	bidirectional switch	(io, io, $\bar{c}$)
tranif1	rtranif1	bidirectional switch	(io, io, c)
tran	rtran	bidirectional buffer	(io, io)
pullup		pull-up	(o)
pulldown		pull-down	(o)

Figure 3.6: Verilog's gate- and switch-level primitives. Inverting variants negate their outputs; resistive variants reduce drive strength from input to output according to the second and third columns of Figure 3.7. The fourth column lists the connections on each primitive: i is an input, o an output, io is both, c is an active-high control signal, and $\bar{c}$ is active-low.

Name	Strength	Reduced	Use
supply	7	5	Power supply
strong	6	5	Gate (default)
pull	5	3	Gate
large	4	2	trireg
weak	3	2	Gate
medium	2	1	trireg (default)
small	1	1	trireg
highz	0	0	

Figure 3.7: Verilog switch-level drive strengths. The third column lists the output strength of a resistive primitive from Figure 3.6 whose input strength appears in column two.

There are five level-sensitive input patterns that can be used in both combinational and sequential UDPs:

Symbol:	0	1	b	x	?
Matches:	0	1	0 1	x z	0 1 x z

Sequential UDPs can also respond to transitions. At most one of these may appear per line, meaning a UDP can only respond to one transition at a time, and cannot only respond to simultaneous ones. The p and n symbols match possible positive and negative transitions.

Symbol:	(vw)	*	r	f	p	n
Matches:	$v \rightarrow w$	(??)	(01)	(10)	(01) (0x) (x1)	(10) (1x) (x0)

The output is set by the symbol at the end of each line. It can be 0, 1, x, or in sequential UDPs, also - (this indicates the output is unchanged). A UDP cannot represent a tristate element.

3.6   Continuous Assignment

Verilog's continuous assignment statements are a shorthand for assembling primitive logic functions within a module. Particularly convenient for arithmetic, a continuous assignment forces the net on its left to be always equal to the expression on its right. During simulation, a change in any of the values appearing in the expression forces it to be reevaluated. User-defined functions may appear in the expression. Output strength and delay may also be specified.

```
wire       carry_out, carry_in;
wire [3:0] sum_out, ina, inb;
assign #15 {carry_out, sum_out} = carry_in + ina + inb;
```

3.7   Parameters and Macros

Verilog's parameters are compile-time constants that can be used to set delays, as symbolic state encodings, and to customize width of busses to provide a primitive form of polymorphism. Each parameter has a name and a default value that may be a constant or a constant expression. Parameter definitions appear within a module body and can be used in expressions. For example:

```
primitive mux2(y, a, b, sel);
  output y;
  input a, b, sel;
  table
//  ab sel y
    1? 0 : 1;
    0? 0 : 0;
    ?1 1 : 1;
    ?0 1 : 0;
    00 ? : 0; // Select unknown but irrelevant
    11 ? : 1;
  endtable
endprimitive

primitive dLatch(q, load, data);
  output q;
  reg q;
  input load, data;
  table
//  c l    st  q
    1 0 : ? : 0; // Latch when clock high
    1 1 : ? : 1;
    0 ? : ? : -; // Hold when clock low
  endtable
endprimitive

primitive dFlipFlop(q, clock, data);
  output q;
  reg q;
  input clock, data;
  table
//  cl da  st  qn
    r  0 : ? : 0; // Load 0 on pos edge
    r  1 : ? : 1; // Load 1 on pos edge
    f  ? : ? : -; // Ignore clock's neg edge
    ?  * : ? : -; // Hold when clock steady
  endtable
endprimitive
```

Figure 3.8: User-defined primitives for a two-input multiplexer, a level-sensitive D latch, and an edge-sensitive D flip-flop.

```
module adder(y, a, b);
  parameter width = 8, delay = 0;
  output [width-1:0] y;
  input [width-1:0]  a, b;
  assign #delay y = a + b;
endmodule
```

Parameter values may be overridden when a module is instantiated. For example, this adder might be instantiated with

```
adder #(4)    (y1, a1, b1);  // width=4
adder #(16,10) (y2, a2, b2);  // width=16, delay=10
```

The infrequently-used defparam directive can cross hierarchical boundaries to override parameters of any instance. This provides a way to consolidate all parameter settings, typically delays, in a single module. For example:

```
module top;
  // ...
  adder adder1(y1, a1, b1), adder2(y2, a2, b2);
endmodule

module annotate;
  defparam
    top.adder1.width = 3,
    top.adder2.delay = 10;
endmodule
```

Verilog has a preprocessor much like C's (see Section 7.8) that allows macros, conditional compilation, and file inclusion. These directives and uses of the macros all start with an accent grave or backquote, written `. Unlike C, directives may appear anywhere, not just in the leftmost column. Typical uses:

```
`define size 8
reg [`size-1:0] a;

`define behavioral 1
`ifdef behavioral
  wire a = b & c;
`else
  and a1(a,b,c);
`endif

`include "myfile.v"
```

3.8   Behavioral Code: Initial and Always blocks

Verilog's initial and always blocks define sequential processes within a module. Such a process runs until it hits a wait or delay statement, causing it to suspend to be reawakened either by the advance of time or an event such as a clock edge. Each process generally has a single control point, although it may internally fork into two or more processes. Statements within a process consist of assignments of expressions' values to registers, conditional and looping constructs, delay and wait statements, fork statements, and others. The one difference between initial and always comes when control reaches the end: an initial terminates, while an always restarts. Thus, initial blocks are used mostly for initialization code and always blocks are used for the real work.

The main reason to execute sequential code is to change the value of registers. Verilog provides two types of assignment that does this. A blocking write (a = b) behaves more like software and immediately writes its register so statements following it will receive the new value. A non-blocking write (a <= b) is more hardware-like and only updates its register after the current simulation timestep (and is thus slower). Nonblocking assignments are useful when a group of signals should all change at some future time:

```
initial begin // Nonblocking
  a <= #10 5; // Happens at time 10
  b <= #15 6; // Happens at time 15
end

initial begin // Blocking
  c = #10 7;  // Happens at time 10
  d = #15 8;  // Happens at time 25 = 10+15
end
```

Non-blocking assignment can also help to avoid nondeterministic race conditions. For example, the pair of processes

```
always @(posedge clock) a = b;
always @(posedge clock) b = a;
```

have a race condition. On a clock edge, these two blocking assignments will run in an undefined order, resulting in both a and b having either the previous value of a or b.

Using nonblocking assignments cures this problem:

```
always @(posedge clock) a <= b;
always @(posedge clock) b <= a;
```

These two will reliably swap the two values. This works because the update events created by the two nonblocking assignments are scheduled separately and run at the "end" of the particular moment in time the clock rises.

Another contrast between the two types of assignment comes when a sequence of them runs. In

```
always @(posedge clock) begin
   b = a;
   c = b;
end
```

c gets the new value of b (i.e., the value of a). By contrast, in

```
always @(posedge clock) begin
   b <= a;
   c <= b;
end
```

c gets the old value of b.

Such behavior is useful for modeling an edge-triggered flip-flop, which effectively latches its data before changing its output.

Assignments may or may not allow other processes to run, which can often appear to violate the idea of a concurrently-running system. For example,

```
module contextswitch;
   integer    q;
   wire [3:0] p;

   assign p = q;

   initial begin
      q = 3;
      $display(p);   // may or may not see 3
   end
endmodule
```

Procedural continuous assignment overrides normal procedural assignment. This can combine asynchronous with synchronous behavior, such as in a flip-flop with an asynchronous clear:

```
module dffc(q, d, clr, clk);
  input  d, clr, clk;
  output q;
  reg    q;

  always @(clr) // Asynchronous clear
    if (clr) #10 assign q = 0; // Override q = #10 d;
    else     #10 desassign q;  // Allow q = #10 d;

  always @(posedge clk) q = #10 d;
endmodule
```

Control-flow statements control the order in which statements are executed within sequential code. Conditional statements, such as if-else and case (see Figure 3.2), run different blocks of code depending on the value of an expression; loops execute blocks repeatedly.

The case statement requires its labels (which may be non-constant expressions) to match exactly. The casez statement relaxes this: a z input will match 0, 1, or x. The casex statement treats both x and z inputs this way.

Verilog provides four types of loops. The forever is an infinite loop, while runs while its expression is true, an expression dictates the number of iterations for a repeat loop, and a for loops specifies initial, repeat, and continue expressions:

```
forever begin body; end
while (i>0) begin i=i-1; end
repeat (j+10) begin body; end
for (i=0; i<10; i=i+1) begin body; end
```

The disable statement can exit named blocks, including loops:

```
begin: breakblk
  forever
    begin: continueblk
      disable continueblk; // restart loop body
      disable breakblk; // exit loop
    end
end
```

Most statements do not advance simulation time, but others can suspend a process and wait for a fixed period of time, wait for events, or wait for a condition. The most common is the event control signal, most often used to wait for the next clock edge (@(posedge clock)), but such an event control statement can also be used to wait for changes in signals or pure events (@(a or b)). A delay such as #10 causes the process to suspend for ten time units. The wait statement can wait for an expression to become true, e.g., wait(a==10).

A module may define named events that can be used to model communication more abstractly. The -> statement generates them and event control statements respond to them.

```
event ready;       // declare the ready event
initial -> ready; // generate the ready event
always @(ready)    // respond to the ready event
  begin /* ... */ end
```

Statements between a fork-join pair run concurrently instead of sequentially. When the fork statement runs, it starts all the processes in the group. The join statement waits for them all to complete before passing control to the statement after it.

The disable statement is often used with fork-join blocks. The fork can be named, allowing a disable statement to terminate all the processes it contains and immediately pass control to the statement after the join.

```
module bigsystem;
  always begin
    reset;                            // Reset task
    fork: mainBody
      forever maintask;               // Main task
      @(posedge reset) disable mainBody; // Terminate main
    join
  end
endmodule
```

3.9   Tasks and Functions

Verilog's tasks and functions are much like procedures and functions in other programming language. They provide a structuring mechanism and can enable reuse. The body of a task or function is sequential code, but since functions must always complete when they are called, they may not contain delay (#10) or event control (@(...)) statements or call tasks. Tasks are called from sequential code; functions may appear in expressions both in sequential code or in a continuous assignment. Tasks take zero or more input, inout, or output arguments. Functions take one or more input arguments and always produce a result.

In the 1995 standard, the registers in tasks and functions are not on a stack, so they may not be called recursively. (A proposed change to the language would add such a stack.)

```
function [8:0] add;  // Return type is nine-bit vector
  input [7:0] a, b;  // Arguments are eight-bit vectors
  begin
   add = a + b; // Return value
  end
endfunction

assign c = add(a,b); // Function call

task multiply;
  input   [31:0] a, b;
  input          clk;
  output [63:0] c;
  begin
   // ...
   @(posedge clk);
   // ...
  end
endtask

initial multiply(a, b, clk, c); // Task invocation
```

3.10   The Programming Language Interface (PLI)

Verilog provides a standard mechanism for linking user-supplied C code to Verilog simulators. This Programming Language Interface (PLI) is a library of C functions that is most often used to write test benches in C when writing a similar model in Verilog is cumbersome or impossible, but can also be used to analyze circuit structure, annotate delay values, interface to actual hardware, and extract and display simulated behavior.

The PLI contains three groups of functions. Task/function routines add system functions (e.g., something like the $monitor system call). Access routines provide the ability to access and modify information (but not topology) in structural Verilog models. Procedural interface routines provide object-oriented access to structural and to behavioral objects.

3.11   Logic Synthesis and the Register-Transfer Level Subset

Verilog was designed as a simulation language, and many of its constructs (e.g., time delays, switch-level semantics) are geared toward what can be simulated rather than what can be synthesized. Nevertheless, Verilog is commonly used as an input language for logic synthesis. Most of the structural and behavioral constructs have fairly obvious translations into gates, and logic synthesis tools are able to automatically translate Verilog into a netlist and optimize it for size and speed.

Instances of gate primitives and continuous assignments are fairly easy to translate, so the main challenge is translating sequential always blocks (initial blocks are not allowed).

The main challenge is interpreting the registers. Depending on how it is used, a reg can represent a wire, a level-sensitive latch, or an edge-triggered flip-flop.

If a register is always assigned when values change, it is combinational. For example, this describes a two-input multiplexer:

```
reg m;
always @(a or b or c)
   if (c == 1)
      m = a;
   else
      m = b;
```

Omitting one of the signals from the event control statement will probably cause a simulation-synthesis mismatch.

However, if m was only assigned when c is true, the result is a level-sensitive latch:

```
reg q;
always @(d or c)
   if (c == 1) q = d; // q not assigned when c is 0
```

Responding to the edge of a signal rather than its level implies edge-triggered flip-flops:

```
red q;
always @(posedge clk)
   q = d;
```

These templates can be extended to much bigger blocks of code. The fsm module in Figure 3.2 can be synthesized.

Translating a Verilog specification into correct gates is easy; transforming the resulting much-to-big-and-slow netlist into something worth building is the job of logic synthesis. Work in the field has been progressing steadily since the 1980s, and there are now several excellent textbooks on the subject, including Hachtel and Somenzi [33], De Micheli [22], and Devadas, Ghosh, and Keutzer [24].

3.12   Event-driven Simulation

Events drive Verilog simulations. A system is a collection of processes that respond to events, typically changes on the values of their inputs. Each event has an integer timestamp and pending events are kept in an ordered event queue. A simulator keeps a single global time and evaluates all events whose timestamp matches it. After this, the simulator advances time to the timestamp of the next earliest event in the queue and the process repeats. Running an process may schedule more events at the current time or in the future, but not in past.

A process is awakened by any event in its sensitivity list, runs, and may generate more events in response. The sensitivity list for a primitive or a continuous assignment is fixed by its inputs, but the sensitivity list of an initial or always block is set by event control, wait, and delay statements and may change with time.

There are two types of event. A change in the value of a net, register, or a named event, is an update event. When an update event runs, it updates the value of its object and generates evaluation events for any processes that are sensitive to the change, e.g., a primitive gate

Current	Future

active

inactive	inactive
nonblocking assign	nonblocking assign
monitor	

While there are events,
    While there are events at the current time,
        If there are any inactive events, activate them,
        else if there are any nonblocking assign events, activate them,
        else activate all monitor events.
        While there are active events,
            Select an active event.
            If it is an update event,
                Update the changed object,
                    scheduling any evaluation events for sensitive processes
            else (it is an evaluation event)
                    Evaluate the process, scheduling any update events.
        Advance simulation time to the next event.

Figure 3.9: Verilog event queue and simulation algorithm

is sensitive to changes on its inputs. Evaluation events run processes, and a running process that writes values generates update events.

Verilog's event model is complicated by nonblocking assignments and monitors, which run after all other events are finished for the current time instant. The semantics are defined operationally in the standard [42, page 46]. The event queue is divided into the six buckets shown in Figure 3.9. Events in the current regions have the timestamp of current simulation time; events in the future region have later timestamps. Simulation proceeds in cycles. In each, events from the next non-empty bucket are activated and evaluated. Only when all the current buckets are empty does simulation time advance.

3.13   Exercises

3–1.   Name three differences between a `reg` and a `wire`.

3–2.   Why do Verilog's literals specify their width along with their value?

3–3.   Give two reasons why you would implement a function with a UDP instead of a continuous assignment or `always` block.

3–4.   A continuous assignment can be implemented as an `always` block. (a) Write an always block for the continuous assignment

```
assign #15 {carry_out, sum_out} = carry_in + ina + inb;
```

(b) Can an always block always be written as a continuous assignment? Why?

3–5.   (a) What can parameters be used for? (b) What common use of parameters can also be done with the preprocessor?

3–6.   Write an `always` block that behaves differently if all its blocking assignments are replaced with nonblocking assignments.

3–7.   Name two sources of nondeterminism in Verilog.

3–8.   Verilog allows registers to be shared between processes. What are the advantages of this? The disadvantages?

3–9.   Why do functions not permit delays (e.g., #3) in their bodies?

3–10.   (Hard) Write a Verilog program that reports how many times a particular continuous assignment statement is evaluated.

4

VHDL

VHDL was designed to be a general modeling and simulation language for digital systems. Its event-driven semantics can be used for everything from switch- to system-level modeling, and it has been pressed into service as an input language for logic synthesis. An analog subset has even been added.

Compared to Verilog VHDL is a big, verbose language (since it was influenced by Ada) with a more sophisticated type system and more mechanisms for structuring large systems. Although VHDL has less built-in for simulating digital circuits (e.g., it has no direct support for four-valued simulation), the language is flexible enough to have later added these facilities as standard libraries.

Like Verilog, VHDL system descriptions may contain a mixture of structural and behavioral modeling styles. VHDL forces the interface of a block (an entity) and its contents (an architecture) to be defined separately. A configuration selects different architectures for an entity.

Figure 4.1 shows the four-NAND XOR circuit written in VHDL. Compare this with Figures 2.5 or 3.1. Both objects are split into two: the entity declaration defines the interface to the object being built, while the architecture declaration defines how it is implemented.

Two of VHDL's three modeling styles are shown here. The architecture for the NAND (arbitrarily named Behavior) illustrates a dataflow style consisting of concurrent assignment statements. The architecture for XOR is structural: four instances of the NAND component are instantiated and connected to signals and ports.

```
entity NAND is
  port (a: in Bit; b: in Bit; y: out Bit);
end NAND;

architecture Behavior of NAND is
begin
  U1: y <= a nand b;
end Structure;

entity XOR is
  port (a: in Bit; b: in Bit; y: out Bit);
end XOR;

architecture Structure of XOR is
  signal i1, i2, i3 : Bit;
  component NAND
    port( a, b: in Bit; y: out Bit );
  end component;
begin
  X1: NAND port map (a, b, i1);
  X2: NAND port map (i1, a, i2);
  X3: NAND port map (i1, b, i3);
  X4: NAND port map (i2, i3, y);
end Structure;
```

Figure 4.1: The XOR build from four NANDs in VHDL

Many books describe VHDL. Perry [67] provides a good tutorial. Bhasker [9] concentrates on the use of VHDL in logic synthesis; Cohen [21] discusses VHDL style issues in the context of logic synthesis. Dewey's book [25] teaches digital engineering using VHDL as a basis. Ashenden's guide for students [4] is a shorter version of his more comprehensive guide [3]. Hunter and Johnson [40] is also targeted toward students. The book edited by Harr and Stanculescu [36] discusses advanced topics such as using VHDL for modeling transmission lines, switch-level behavior, and analog interfaces. The tediously precise IEEE standard [41] is the definitive reference on the language.

4.1 Entities and Architectures

A VHDL system description consists of entities (functional blocks with well-defined inputs and outputs) and architectures that define the contents of an entity. An architecture may include processes, concurrent signal assignments, and instances of other entities. Multiple architectures for the same entity may be defined and later selected by a configuration. This level of indirection can be used, say, to replace a procedural description of a datapath with one built from gates. Think of an architecture as a chip, an entity as a packaged chip, a component as a socket for the chip, and a configuration as a parts lists that specifies which chip to insert in each socket.

The most basic entity is a named list of ports:

```
entity mux2 is
  port( a, b, c: in Bit; d: out Bit);
end;
```

Ports may be inputs, outputs, inout (a bidirectional port), buffer (a bidirectional port isolated from external drivers), or linkage (for documentation).

Entities may also define generics: compile-time constants used to define such things as bus widths and delays. They may include default values:

```
entity adder is
  generic (W: Natural := 31);
  port (a, b: in  Bit_Vector (W downto 0);
        s   : out Bit_Vector (W downto 0); );
end adder;
```

Entities are designed to contain things that are independent of implementation, such as the interface, generics, and monitors that verify the correct behavior of the environment or implementation. For maximum flexibility, entities may declare types, constants, local signals, package use statements, assignments, and processes.

An entity may have one or more associated architectures. Each consists of a name, declarations, and a body that can include instances, concurrent signal assignments, and sequential processes. Here are examples of three styles of architecture body:

```
--------------------------------------- STRUCTURAL STYLE
architecture arch1 of mux2 is

  signal cbar, ai, bi : Bit;              -- internal signals

  component Inverter                      -- component interfaces
    port (a:in Bit; y: out Bit);
  end component;

  component AndGate
    port (a1, a2:in Bit; y: out Bit);
  end component;

  component OrGate
    port (a1, a2:in Bit; y: out Bit);
  end component;

begin
  I1: Inverter port map(c => a, y => cbar); -- connect-by-name
  A1: AndGate  port map(a, c, ai);        -- connect-by-position
  A2: AndGate  port map(a1 => b, a2 => cbar, y => bi);
  O1: OrGate   port map(a1 => ai, a2 => bi, y => d);
end;

--------------------------------------- DATAFLOW STYLE
architecture arch2 of mux2 is
  signal cbar, ai, bi : Bit;
begin
  cbar <= not c;
  ai   <= a and c;
  bi   <= b and cbar;
  d    <= ai or bi;
end;

--------------------------------------- BEHAVIORAL STYLE
architecture arch3 of mux2 is
begin
  process(a, b, c)  -- define sensitivity list
  begin
    if c = '1' then
      d <= a;
    else
      d <= b;
    end if;
  end process;
end;
```

4.2   Structural Description

Structural description is hierarchical: an architecture body written in a structural style instantiates components and connects them, just like any netlist format. Thus, the component instantiation statement is the main construct, and connecting the ports on that component to the signals and ports of the entity it is contained in is the main detail.

Each port on an entity is a signal with a name and direction. These are declared with the `port` directive. For example, a seven-segment decoder entity might be defined as:

```
entity DecoderSeven is
  port (enable: Bit; a: in Bit_Vector (3 downto 0);
        segments: out Bit_Vector (6 downto 0);)
end DecoderSeven;
```

Each component instantiated within an architecture must be defined within a `used` package or in the architecture. This extra level of indirection allows different architectures or entities to be associated with the same component.

Internal signals are declared with the `signal` directive in the declaration section (before the architecture's `begin`).

Component instantiation statements have an instance name, component name, and a port map listing how the component's ports are connected to signals and ports within the architecture. These connections can be by position or by name.

```
architecture arch1 of DecoderSeven is
  component And2
    port(a1, a2: Bit; y: out Bit);
  end component;
  signal int: Bit_Vector(6 downto 0);
begin
  and_1: And2 port map(enable, a(0), int(0));
  and_2: And2 port map(a1 => enable, a2 => a(1), y => int(1));
end arch1;
```

An output port connected to "open" or not listed is unconnected:

```
  and_3 : And2 port map(a, b, open);
  and_4 : And2 port map(a1 => a, a2 => b);
```

If an entity has generics (compile-time constants), they are defined along with its ports. An ALU might have a generic for the widths of its data paths:

```
entity ALU is
  generic (Width: Positive);
  port (arg1, arg2: Bit_Vector (Width-1 downto 0);
        out: out Bit_Vector (Width+2 downto 0); );
end ALU;
```

Such a component is instantiated with port and generic maps:

```
architecture ex of UsesAlu is
  component ALU
    generic (Width: Positive);
    port (arg1, arg2: Bit_Vector (Width-1 downto 0);
          out: out Bit_Vector (Width+2 downto 0););
  end component;
  signal a1, a2: Bit_Vector (31 downto 0);
  signal r: Bit_Vector (33 downto 0);
begin
  Alu1: ALU
    generic map (Width => 32)
    port map (a1, a2, r);
end
```

A configuration specification defines which entities and architectures each component uses. It can remap ports on an entity to match those on the component instantiating the entity. For example, to select between two architectures:

```
entity foo is
  port( fa: Bit, fb: out Bit);
end;
architecture arch1 of foo is ... end;
architecture arch2 of foo is ... end;

architecture config_ex of ex is
  component bar port( ba: Bit, bb: out Bit); end component;
  for bar1: bar use entity Work.foo(arch1)
    port map(fa => ba, fb => bb);   -- remap the ports
  signal a, b: Bit;
begin
  bar1: bar port map(a, b);
end;
```

Here, Work.foo indicates entity foo in library Work, where the foo entity was placed by default.

The generate statement provides a way to regularly instantiate components using simple loops and conditionals. As a contrived example, consider this array of gates to AND the bits of a vector:

```
entity wideand is
  generic (Width: Positive);
  port(i: Bit_Vector(Width-1 downto 0); o: out Bit;);
end wideand;

architecture gen_ex of wideand is
  component and2 is
    port(A1, A2: Bit, Y: out Bit);
  end component;
  signal t: Bit_Vector(Width-2 downto 0);
begin
  for K in 1 to Width-1 generate

    if K = 1 generate
      A: and2 port map(i(0), i(1), t(0));
    end generate;

    if K > 1 and K < Width-1 generate
      B: and2 port map(t(K-1), i(K), t(K));
    end generate;

    if K = Width-1 generate
      C: and2 port map(t(K-1), i(K), o);
    end generate;

  end generate;
end gen_ex;
```

4.3   Dataflow Description: Concurrent Signal Assignment

VHDL allows architectures to contain concurrent signal assignment statements that behave like combinational logic that continually computes a function of its inputs. The expression driving a signal may use the operators in Figure 4.2 as well as conditionals and delays.

One of the two types of concurrent signal assignment allows an expression to select the reaction. For example:

```
architecture concurrent of ex is
begin
  Inc <= 0       after 10 ns when Enable = '0' else
         In1+1 after 50 ns when Sel = '1'    else
         In1   after 10, In1+1 after 40, In1 after 50;

  with Op select      -- selected concurrent signal assignment
    Result <= In1 or  In2 after 20 ns when "000",
              In1 and In2 after 20 ns when "001",
              In1 + In2   after 50 ns when "010",
              In1 - In2   after 60 ns when "011",
              0           after 10 ns when others;
end
```

Concurrent signal assignment statements are equivalent to processes sensitive to the signals read by the expression.

4.4 Behavioral Description: Processes

The main behavioral construct is the process: a group of sequential statements executed in order (like a C function). Processes run concurrently when they are not suspended. Sequential statements include variable and signal assignments, control-flow instructions such as if-then-else, and wait statements that suspend the process and control what will reawaken it. The body of each process behaves as if it were enclosed in an infinite loop, so at least one wait statement (or a shorthand for it) per process is compulsory.

A process always resides within an architecture body and may contain declarations for local variables, constants, and signals.

```
architecture behavior of ex1 is
begin
  Counter: process                  -- name is optional
    variable count: Bit_Vector (3 downto 0) := "0000";
  begin
    count := count + 1;            -- variable assignment
    if count = 10 then             -- conditional
      count := 0;
    end if;
    countout <= count after 5ns;   -- signal assignment
    wait until Clk'EVENT and Clk = '1'; -- wait for pos. edge
  end process;
end;
```

Variables defined within a process hold data exclusively for that process and cannot be shared with others. This avoids race conditions with shared variable access. (Verilog permits this source of nondeterminism; Java uses semaphores to avoid shared variable contention. See Section 9.6 on Page 172.)

Newer versions of VHDL allow variables to be defined within an architecture and shared among its processes with the usual caveat that this can produce nondeterministic behavior.

VHDL uses := for variable assignment. The variable to its left immediately gets the value of the expression on its right. For example,

```
count := count + 1;
V := X;
```

The most basic form of signal assignment assigns a value to a signal after the process has suspended:

```
OutSignal <= '1';
```

This means the process cannot immediately read the newly-written value:

```
ValueSignal <= 10;
a := ValueSignal;   -- a gets the earlier value, not 10
```

The after directive delays the assignment further:

```
OutSig <= '0' after 100 ps;
```

VHDL can only model causal systems: the delay must always be nonnegative. Only the future can be changed, not the past.

A signal assignment sets the latest event on a signal, causing any later events to be discarded. For example,

```
signal S : integer := 0;
P1: process begin
  S <= 1 after 2 ns;
  S <= 2 after 1 ns; -- Discard event at 2 ns
  wait;              -- terminate
end process;
```

By default, a delay in VHDL models devices that ignore glitches. Scheduling a later event removes earlier events:

```
signal S1, S2 : integer := 0;
P1 : process begin
  S1 <= 1 after 1 ns;
  S1 <= 2 after 2 ns; -- Discard event at 1 ns
  S2 <= 2 after 1 ns;
  S2 <= 2 after 2 ns; -- Event at 1 ns unchanged
  wait;               -- terminate
end process;
```

Transport delay is an alternative to the default (inertial) that models situations where brief changes in value are not ignored, such as transmission lines. A transport assignment preserves earlier events, only discarding later ones:

```
signal S3, S4 : integer := 0;
P2 : process begin
  S1 <= transport 1 after 2 ns;
  S1 <= transport 2 after 1 ns; -- Discard event at 2 ns
  S2 <= transport 1 after 1 ns;
  S2 <= transport 2 after 2 ns; -- Event at 1 ns unchanged
  wait;
end process
```

Multiple events can be scheduled on the same signal to produce a waveform such as a clock:

```
Clock <= '1' after 100ns, '0' after 200ns,
         '1' after 300ns, '0' after 400ns;
```

To make this behave as expected, the second and later events in such an assignment are treated as having transport delay.

Because all signal assignment happens in the next simulation cycle, the effect of signal assignment is not available immediately:

```
signal A : Bit := '0';
signal B : Bit;

P1 : process begin
  A <= '1';
  B <= A;   -- B <= '0' because A was not updated
end process;
```

Inserting a wait statement avoids this problem by allowing the update to occur:

```
P2: process begin
  A <= '1';
  wait for 0ns;    -- force the assignment to occur
  B <= A;          -- B <= '1'
end process;
```

VHDL provides classical control constructs such as if-then-else, a multi-way case statement, while and infinite loops, and the exit and next statements for exiting and restarting loops.

The if-then-else statement has three forms:

```
if a = 10 then       -- if-then
  b := 20;
end if;

if count < 20 then  -- if-then-else
  c := 30;
else
  c := 40;
end if;

if b = 0 then        -- if-then-elsif-else
  c := c + 1;
elsif b = 1 then
  c := c + 2;
else
  c := c + 3;
end if;
```

The case statement provides multi-way decisions:

```
case state is             -- expression
  when IDLE =>            -- case label must be a constant
    if A = '1' then       -- sequence of statements follows
      nextState := RUN;
    end if;
    running := '1';

  when RUN | RUN2 =>      -- multiple choices in a rule
    if A = '1' then
      nextState := IDLE;
    end if;

  when others =>          -- default rule
      error := '1';
end case;
```

VHDL provides conditional and infinite loops. Loops may be labeled to provide a named target for the `exit` and `next` statements, which terminate and restart loops respectively.

```
Outer: loop                   -- infinite loop
  i := i + 1;
  if i = j then
     next Outer;              -- restart this loop
  if count = 10 then
     exit;                    -- terminate this loop
  exit Outer when i = k;  -- conditional termination
end loop;

while i < 10 loop             -- while loop
  i := i + 1;
  j := j + 10;
end loop;

for i in 1 to 10 loop      -- for loop
  a(i) := a(i) + i;
end loop;
```

A process runs until it reaches a `wait` statement, which suspends the process and defines what will reactivate it. A process is reawakened by some signal changing when a condition is true or after a specified timeout. Any or all of these clauses may be omitted.

```
wait;                              -- terminate the process
wait on A, B;                      -- restart when A or B changes
wait on D until En = '1';     --   when D changes and En is 1
wait on Clk until Clk = '1'; --   when Clk rises
wait for 10 ns;                --   in 10 ns
wait on A for 10 ns;           --   when A changes or in 10 ns
wait on Clk until Clk = '1' for 5ns;
```

Combinational logic is most naturally modeled with a process that is always sensitive to changes on the same set of signals. VHDL provides a shorthand:

```
process(A, B, C) begin
  ...
end process;
```

is equivalent to

```
process begin
  ...
  wait on A, B, C;
end process;
```

The `assert-report` statement checks a condition and flags an error if it is false. This is useful for debugging:

```
process
begin
  c := a + 3;
  assert c < 5 report "c is out of range" severity Error;
end process;
```

Although `assert` statements can be placed within an architecture, placing them in entities is convenient for checking I/O behavior independently of the entity's implementation.

4.5   Procedures and Functions

VHDL systems may encapsulate sequential code in functions and procedures much like their software language counterparts. A function computes a return value and may not modify any of its arguments, nor invoke `wait`. In contrast, a procedure is only useful for its side effects: it does not return a value, but has inputs, outputs, and inout parameters, and may call `wait`. A function is assumed to be pure by default, indicating it will always return the same value given the same arguments. If this is not the case, it should be marked `impure`.

Functions and procedures may be declared separately from their definition. A declaration defines the interface (parameters and the return type of a function), while a definition also defines the body of the subprogram.

```
function Min(constant X, Y: Integer) return Integer;    -- decl
function Min(constant X, Y: Integer) return Integer is -- def
begin
  if X < Y then
    return X;
  else
    return Y;
  end if;
end Min;
```

```
procedure Swap(X, Y: inout Integer) is
  variable Temp : Integer;
begin
  Temp := X;
  Y := X;
  X := Temp;
end;


procedure Alu(Arg1, Arg2: in  Integer;
              Op         : in  Operation := NoOp; -- default
              Res        : out Integer);
```

A procedure call is a statement, a function call is an expression:

```
Swap(a, b);                                     -- procedure call
c := Min(d, e);                                 -- function call
Alu(Arg1 => x1, Arg2 => x2, Res => result); -- procedure call
```

VHDL signals normally have a single driver, but VHDL also allows resolved signals that have multiple drivers. Each resolved signal has a resolution function that considers all driver values to resolve the value of the signal. Resolved signals are typically used to model wired-AND and wired-OR configurations.

A resolution function takes a one-dimensional unconstrained array and produces a single value. For example,

```
type FourV  is ('0', '1', 'X', 'Z');
type FourVV is array (Integer range <>) of FourV;


function WiredAnd(Values: FourVV) return FourV is
  variable Result : FourV := '1';
begin
  for I in Values'Range loop
    if Values(I) = '0' then
      Result := '0';
    elsif Values(I) = 'X' then
      Result := 'X';
      exit;
    end if;
  end loop;
  return Result;
end WiredAnd;


signal ResolvedSignal : WiredAnd FourV;
```

4.6   Types and Expressions

VHDL sports a powerful type system borrowed from Ada that provides enumerated types, numbers with ranges, physical types such as time, arrays, and records with named fields. New types may be built by subclassing other types (e.g., the literals of one enumerated type may be a subset of another).

VHDL provides four scalar types: integers, floating-point numbers, enumerated types, and physical quantities. Type definitions for integers and floating-point numbers include a range specification. These ranges constrain the values a scalar value may take.

```
type date is range 1 to 31;
type address is range 16#0000# to 16#FFFF#;
```

Literal numbers are interpreted in base ten by default, but can be specified in any base between two and sixteen. There are two notations: 16#C080# and H"C080" represent the same integer. Floating-point literals contain a decimal point: 10e4 is an integer; 10.0e4 is floating-point. Such literals may also be specified in a different base such as octal: 8#43.6#e+4. Especially useful for long binary numbers, underscores within a lieral are ignored: B"1010_1101".

The literals in an enumerated type may be single characters or identifiers. Enumerated identifiers may be reused: context generally resolves ambiguities.

```
type Bit is ('0', '1'); -- A built-in type
type State is (HighGreen, HighYellow, FarmGreen, FarmYellow);
type OtherState is (High_Green, High_Yellow, NotGreenYellow);
```

Physical types are comprised of a range constraint, a base unit, and zero or more secondary units that are integer multiples of the base unit. A literal for a physical type includes its units, e.g., 10 ms.

```
type Time is range -(2**31-1) to (2**31-1)   -- Built-in
  units
    fs; ps = 1000 fs; ns = 1000 ps; us = 1000 ns;
    ms = 1000 ns; sec = 1000 ms; min = 60 sec; hr = 60 min;
  end units;
```

VHDL provides two composite types: records and arrays. Records contain named fields, each of which can be a different type. Arrays contain objects of the same type that can be indexed with integers or enumerated types. Arrays may have one or more dimensions. Dimensions of an array may be constrained or unconstrained—able to grow during the simulation.

```
type Opcode      is (mov, add, adc, sub, sbc);
type Register    is range 0 to 15;
type Instruction is record
  Op               : Opcode;
  Src1, Src2, Dest: Resister;
end record;
type Byte     is range 0 to 255;
type Address  is range 0 to 16#ffff#;
type Memory   is array (Address)       of Byte; --Constrained
type Img      is array
  (Integer range <>, Integer range <>) of Byte; --Unconstrained
```

Fields of records are accessed with the dot: inst.Op. Elements of an array are accessed with parenthesis: img(100,150).

Strings are actually a built-in array type, part of the standard package (Figure 4.5):

```
type STRING is array (Positive range <>) of CHARACTER;
```

String literals are surrounded by double quotes ("Hello world") and doubled double quotes become a single one in the string, e.g., """Hello,"" she said.".

Subtype declares new types by constraining existing ones:

```
type Address      is range 0 to 16#ffff#;
subtype IOAddress is Address range 16#c000# to 16#c800#;
type Opcode       is (mov, add, adc, sub, sbc);

subtype ArithOp   is Opcode add to sbc;
subtype Op        is Opcode; -- Op is an alias for Opcode
```

VHDL's expressions are built from literals and the operators listed in Figure 4.2.

The shift, rotate, and logical operators operate on the predefined types Bit and Boolean or one-dimensional arrays of them. VHDL expects busses and vector-valued registers to be represented this way.

n ** i	exponentiation	abs n	absolute value
not b	logical negation		
n * n	multiplication	n / n	division
i mod i	modulus	i rem i	remainder
+n	(unary)	-n	negation
n + n	addition	n - n	subtraction
a & a	concatenation		
b sll i	shift left logical	b srl i	right logical
b sla i	shift left arithmetic	b sra i	right arithmetic
b rol i	rotate left	b ror i	right
x = x	equal	x /= x	not equal
n < n	less than	n > n	greater than
n <= n	... or equal	n >= n	... or equal
b and b	logical AND	b nand b	logical NAND
b or b	logical OR	b nor b	logical NOR
b xor b	logical XOR	b xnor b	logical XNOR

n:	numeric type	i:	integer
x:	any non-file type	b:	scalar or vector Bit or Boolean
a:	one-dimensional array		

Figure 4.2: VHDL operators in groups of decreasing precedence.

The concatenation operator combines one-dimensional arrays (or scalars that compose such arrays) to form wider arrays. This operator reverses the effects of slicing arrays:

```
type Word is array(15 downto 0) of Bit;
variable input, output : Word;

output := input(15 downto 8) & input(0 to 7); -- reverse lsb
```

The modulus and remainder operators only differ given negative operands. The remainder has the sign of its dividend; the modulus has the sign of its divisor. In both cases, the absolute value of the remainder or modulus is less than the absolute value of the divisor.

The predefined operators may be overloaded. The standard package IEEE.Std_logic_1164 overloads them for multi-valued logic types.

4.7   Attributes

Most VHDL objects, such as types, functions, variables, and packages, have attributes: named characteristics of an object that are separate from its value. Attributes are accessed with the ' operator. For example, a range type has attributes that report its lowest and highest value:

```
type BitIndex is range 31 downto 0;

-- BitIndex'Left is 31
-- BitIndex'Right is 0
-- BitIndex'Low is 0
-- BitIndex'High is 31
```

Many objects have such predefined attributes, but VHDL also permits user-defined attributes. The type of an attribute must be defined before the attribute can be associated with an entity:

```
attribute NumberOfPins : Integer;
attribute NumberOfPins of NANDChip is 14;
```

VHDL's attribute mechanism provides a powerful introspection capability and provides access to simulation behavior. The six classes

General type attribute

'BASE	base type of derived type

Scalar type attributes

'LEFT 'RIGHT	left, right bound
'LOW 'HIGH	lower and upper upper bound
'ASCENDING	true when signal has ascending range
'IMAGE(x)	printable representation of x
'VALUE(s)	interpretation of string s

Enumerated type attributes

'POS(x) 'VAL(i)	integer position of x, value at position i
'PRED(x) 'SUCC(x)	one lower, higher position than x
'LEFTOF(x) 'RIGHTOF(x)	one left, right than x

Array attributes

'LEFT(d) 'RIGHT(d)	left, right bound
'LOW(d) 'HIGH(d)	lower, upper bound
'RANGE(d)	range
'REVERSE_RANGE(d)	reversed range of array dimension d
'LENGTH(d)	length of array dimension d
'ASCENDING(d)	true if array dimension d is ascending

Signal attributes

'DELAYED(t)	signal: the signal delayed by t
'STABLE(t) 'QUIET(t)	signal: true if no events, no activity for t
'TRANSACTION	signal: toggles on activity
'EVENT 'ACTIVE	true on event, activity
'LAST_EVENT	elapsed time since last event
'LAST_ACTIVE	elapsed time since last activity
'LAST_VALUE	signal state before last event
'DRIVING	true if being driven
'DRIVING_VALUE	value being driven

Named entity attributes

'SIMPLE_NAME	the simplified string name of the entity
'INSTANCE_NAME	hierarchical name of the entity
'PATH_NAME	hierarchical name excluding the instance

Figure 4.3: Predefined attributes. d represents an optional index of a array's dimension (1 is first). s represents a string. i represents an integer. t represents an optional time.

of predefined attributes, listed in Figure 4.3, provide queries about scalar, enumerated, and array types, along with information about signal behavior and named entities.

The IMAGE and VALUE attributes are functions that convert scalar values to and from string representations.

The enumerated type attributes allow queries about the order of entries in the type, along with ways to step through them.

The array attributes return information about the dimensions of array types. All have an optional dimension index parameter that, when omitted, defaults to one.

The attributes of signals are the most powerful. The DELAYED attribute returns a version of the signal delayed by the given amount of time. The remaining signal attributes draw a distinction between when a signal is active, indicating something has assigned a value to it, and has an event, indicating activity has changed its value. Electrical simulations are generally event-driven, meaning their components change only when an input does, but higher-level models often use signals to model communication channels that respond to communication, rather than just value changes.

The named entity attributes provide a way to report on the structure of entities in the model.

4.8   Packages and Libraries

A package contains declarations (e.g., types, procedures) that can be shared across entities. For example, the STANDARD package defines built-in types such as Bit and Boolean, and the TEXTIO package defines functions for printing text to files.

Normally, types, constants, and subprograms defined within an entity are invisible outside it. A VHDL package groups such definitions so they can be used by more than one entity. For example, an instruction package might contain definitions used by parts of a processor responsible for decoding and executing instructions. Like entities, packages are broken into an interface and a body.

```
package Instruction is
   type Opcode is (ADD, ADC, SUB, SBC, MOV);
   type Register is range 0 to 15;
   type Instruction is range 0 to 16#ffff#;
```

```
   type DecodedInst is record
     Op: Opcode;
     Lsrc, Rsrc, Dest: Register;
   end record;
   function Decode( In: Instruction ) return DecodedInst;
end Instruction;

package body Instruction is
   function Decode( In: Instruction ) return DecodedInst is
   begin
     -- ...
   end Decode;
end Instruction;
```

The use directive makes declarations within a package visible to the entity or architecture definition that immediately follows it:

```
use Instruction.all;
architecture arch1 of myDecoder is
   signal r: Register;
   siglal op: Opcode;
begin
   ...
end
```

A library is an implementation-dependent container for entities, architectures, packages, and all other design objects. The VHDL standard does not dictate how to create and define libraries, but it does say how to use them and defines some standard libraries.

The library directive makes a library visible so that it can be used. For example, the following brings the IEEE library into scope and then makes available all the declarations in the Std_logic_1164 package:

```
library IEEE;
   use IEEE.Std_logic_1164.all;
```

By default, VHDL assumes each file begins with

```
library STD, WORK;       -- Make STD and WORK libraries visible
use STD.STANDARD.all;    -- Make standard package visible
```

Figure 4.5 is a summary of the standard package, which defines VHDL's built-in types.

use Std.Standard.all;	Built-in types
use Std.Textio.all;	Text I/O
use IEEE.Std_Logic_1164.all;	Multi-valued logic
use IEEE.Numeric_std.all;	Signed and unsigned arithmetic

Figure 4.4: Standard VHDL packages

```
package STANDARD is

    type SEVERITY_LEVEL is (NOTE, WARNING, ERROR, FAILURE);
    type BOOLEAN       is (FALSE, TRUE);
    type BIT           is ('0','1');
    type BIT_VECTOR    is array (NATURAL range <>) of BIT;
    type INTEGER       is range ...; -- implementation defined
    subtype NATURAL    is INTEGER range 0 to INTEGER'HIGH;
    subtype POSITIVE   is INTEGER range 1 to INTEGER'HIGH;

    type CHARACTER     is ( -- 256 characters total
        NUL, ..., USP,
        ' ', '!', ..., '?',
        '@', 'A', ..., 'Z', ...
        '~', 'a', ..., 'z', ... );
    type STRING        is array (POSITIVE range <>) of CHARACTER;

    type REAL          is range ...; -- implementation defined
    type TIME          is range ...  -- implementation defined
      units
        fs; ps = 1000 fs; ... hr = 60 min;
    end units;
    subtype DELAY_LENGTH is TIME range 0 fs to TIME'HIGH;
    impure function NOW return DELAY_LENGTH;

    type FILE_OPEN_KIND is (READ_MODE, WRITE_MODE, APPEND_MODE);
    type FILE_OPEN_STATUS is
        (OPEN_OK, STATUS_ERROR, NAME_ERROR, MODE_ERROR);

    attribute FOREIGN: STRING;

end STANDARD;
```

Figure 4.5: The VHDL standard package, which defines built-in types. Each ... indicates omitted details. From IEEE Std. 1076-1993. Copyright 1994 IEEE. All rights reserved.

4.9   Exercises

4–1.   Why does VHDL define components? Why not allow entities to be instantiated directly?

4–2.   Why would you create multiple architectures for the same entity?

4–3.   Write a process equivalent to this concurrent signal assignment:

```
with Op select
Out <= In1 + In2 when "01",
       In1 - In2 when "10",
       0         when others;
```

4–4.   Why did VHDL choose inertial delay as the default instead of transport?

4–5.   What are the three conditions of a wait statement?

4–6.   What is the difference between a pure and impure function? Why would a simulator care?

4–7.   (a) What is a resolution function good for modeling? (b) Why should a resolution function return a value that is independent of the order of the entries in the array it is given?

4–8.   What is the difference between an active signal and one that has an event?

Part II

Software

A software language describes sequences of instructions that a processor will execute. Since each instruction does little (e.g., read two numbers from memory, add them, and write the result), a high-level language aims to specify many of them concisely and intuitively. Variables are named memory locations that can be used in arithmetic expressions such as $ax+by+c$. Groups of expressions in control-flow statements are executed once or more depending on conditions. A function is a group of statements that may be called from an expression. A called function is given control, computes a results, and returns control to the expression that called it. Object-oriented languages create new data types by specifying their contents and functions that may transform them.

5

Software Basics

Software languages describe instructions that will be executed in sequence by a processor. These instructions perform arithmetic on values in the processor's memory or make a decision based on the results. Combinations of arithmetic instructions can be assembled to perform more complex tasks, such as evaluating a polynomial like $ax^2 + bx + c$. Adding decisions makes possible algorithms such as Euclid's for computing the greatest common divisor (Figure 5.1).

Assembly language (described in Chapter 6) provides the most precise control over a processor's instructions and in theory allows a designer to create the smallest, fastest programs possible. But control comes at the cost of having to do it: the level of detail makes writing assembly language slow and error-prone. Nevertheless, designers often resort to assembly language when they need to squeeze the last few cycles out of a crucial inner loop or directly access hardware.

Higher-level languages such as C, C++, and Java give programmers a more succinct, less error-prone notation for common programming idioms including expressions, loops, functions, and data structures.

1: Let r be the remainder of $m \div n$.
2: If $r = 0$, then go to step 6.
3: Set m to n.
4: Set n to r.
5: Go back to step 1.
6: Report n as the result and quit.

Figure 5.1: Euclid's algorithm for computing the greatest common divisor of two positive integers m and n.

The disadvantage comes from the sometimes clunky way the compiler translates the designer's intention into assembly-language instructions. The result can be bigger and slower than what a skilled assembly-language programmer could produce, but for modern 16- and 32-bit general-purpose processors, compilers can generate code nearly as efficient as code produced by a careful person writing in assembly language, and do it much more quickly, making these high-level languages the choice for creating most large software systems. Sadly, the simpler DSPs and microcontrollers (four- and eight-bit) have assembly languages too irregular for most compilers, necessitating manual assembly language coding for efficient code.

Because it generally interacts directly with hardware peripherals and the physical world, software for embedded systems often uses lower-level constructs than software for general-purpose computing. In a general-purpose computing environment, a programmer is usually supplied with an operating system responsible for communicating with I/O devices such as keyboards, disk drives, and video displays. Although the more complicated embedded systems often have such operating systems, it is much more likely that the embedded system designer is at least partially responsible for designing the os.

5.1   Representing Numbers

Integers, specifically integers with fixed ranges, are the primary type of data manipulated by most processors. Such integers are useful for counting things such as iterations of a loop or members of a datastructure, and have the advantage of exact addition and subtraction, provided they don't overflow.

In hardware, integers are represented in binary (base two), so an eight-bit integer can represent the numbers 0 to 255. In general, an n-bit integer can represent numbers from 0 to $2^n - 1$. Negative integers are represented using Two's complement: a binary number where the weight of the most significant digit is negative, i.e., -2^{n-1}. Thus,

$$1101_2 = -1 \cdot 2^3 + 1 \cdot 2^2 + 0 \cdot 2^1 + 1 \cdot 2^0 = -8 + 4 + 1 = -3_{10}$$

Two's complement numbers are added and subtracted with the the same algorithms as binary numbers.

Some applications need to manipulate fractional numbers. A common solution is to use fixed-point numbers: binary numbers whose "decimal point" is fixed somewhere in the middle of the number rather than immediately to the right of the least-significant bit. Fixed-point numbers can be added and subtracted using the same algorithm as for binary integers, but multiplication and division require the result to be shifted.

A typical fixed-point number representation is Two's complement with a decimal point just to the right of the most significant digit, so

$$1.011_2 = -1 \cdot 2^0 + 0 \cdot 2^{-1} + 1 \cdot 2^{-2} + 1 \cdot 2^{-3} = -1 + \frac{1}{4} + \frac{1}{8} = -0.625$$

Like integers, fixed-point addition and subtraction is exact provided it does not overflow.

Fixed-point numbers are poor for representing very small or very large numbers, so floating-point numbers are often provided as an alternative. Floating-point resembles scientific notation,

$$\pm \text{significand} \times 2^{\text{exponent}}$$

representing a number as the product of a sign bit, a significand (a fixed-point number between 1 and 2), and a power of two. The significand carries the "significant digits" to maintain a consistent percentage error across numbers of widely varying magnitude, while the exponent provides wide dynamic range.

In 1985, the IEEE 754 floating-point standard defined a common representation and behavior for floating-point numbers. This standard defined the many corner cases and ambiguous situations that are much more common with floating-point arithmetic than integers.

A number represented in IEEE 754 format consists of a sign bit (1 is negative), the integer exponent to which a fixed offset is added, and the significand in fixed-point format with an implicit 1 just to its left. This implicit 1 makes the value of the significand fall between 1 and 2. For example, the single-precision number

$$1 \quad \underbrace{10000001}_{\text{8-bit exponent}} \quad \underbrace{01100000000000000000000}_{\text{23-bit significand}}$$

$$= \quad -1.011_2 \times 2^{129-127} = -1.375 \times 4 = -5.5.$$

	Bytes	Exponent bits	Significand bits	Significant digits
Single Precision	4	8	24	6–9
Double Precision	8	11	53	15–17
Double-extended	$\geq$ 10	$\geq$ 15	$\geq$ 64	$\geq$ 18–12
Quad Precision	16	15	113	33-36

Figure 5.2: Sizes of IEEE 754 floating-point numbers. Most processors support single and double precision. The others are less common.

Floating-point arithmetic has many more exceptional cases than integer arithmetic, including overflow, underflow, division by zero, and generally unrepresentable results. To handle this, IEEE 754 includes representations of $\pm\infty$ (produced by, e.g., dividing by zero) and NaN (not a number—produced by, e.g., $\sqrt{-1}$).

Floating-point arithmetic is typical in scientific applications such as weather simulations, but it is less frequently used in embedded systems because the hardware is large and power-hungry compared to its integer counterpart, and simulating its behavior in software is usually prohibitively slow. Most embedded applications instead use fixed-point arithmetic after extensive analysis.

Goldberg [37, Appendix A] provides a more through introduction to computer arithmetic.

5.2   Types

A type system defines how pieces of data are represented in memory and restricts the operations that can be performed on them. For example, most type systems draw a distinction between a Two's complement binary number and a floating-point number. This has many advantages. Mainly, it prevents the programmer from inadvertently interpreting one type of bit pattern as another; something almost guaranteed to give a nonsensical result. In addition, it allows data to be represented in different ways and allows representation-specific operations, which are usually much faster.

Some languages are typeless, meaning all operations apply to all objects equally. This eliminates having to convert between representations, but it is usually slower. For example, John Ousterhout's Tool Command Language (Tcl) [64] once represented all types (numbers, lists, and strings), as strings. To add two integers, the Tcl interpreter converted their string representations to binary, added the binary numbers, and converted the result back into a string. Conversion could easily take 100 times longer than the addition itself. For a scripting language such as Tcl, the overhead was tolerable because of the flexibility it provided, but this is unacceptable for most systems language. Later versions of Tcl converted representations lazily: an object represented as an integer would remain an integer until its string representation was needed, at which point it would be converted and stay that way until assigned to another integer.

The BCPL language (a precursor of C—see Section 7.9) is also typeless and uses machine words as its fundamental type. Although this is faster than using strings, the programmer must ensure that string data is never used as a memory address.

The three high-level languages presented in this book, C, C++, and Java, use increasingly sophisticated type systems. C's main focus is on physical types, i.e., how data is represented in memory and does not restrict the operations that can be performed on that data beyond, e.g., preventing integers from being interpreted as floating-point numbers. C++ and Java's type systems add the ability to further restrict the operations on a collection of data through the notion of a class. A class defines both how the data is represented (e.g., integers, strings, etc.) and the operations that can be performed on it in the form of functions or methods. The rest of the program can only access the data through these methods.

5.3   Control Flow

Although some useful computation can be carried out by performing a simple sequence of instructions, most interesting algorithms (such as Figure 5.1) have more complicated control flow, including loops that repeat a section of code (steps 1–5 of Euclid's algorithm), conditionals that run different sections of code depending on intermediate results (step 2), and functions that allows a segment of code to be reused throughout a program without being copied. Euclid's algorithm might be such a function: a bigger program might need to compute greatest common divisors in different places in the program.

In general, a processor can send the point of control anywhere it wants, but such unstructured flow often makes code difficult to understand and debug. Instead, high-level languages encourage structured control flow, which restricts structures to loops and conditionals composed of a "then" and an "else" block. This is a restriction—algorithms are often more elegantly expressed using other structures, but structured control flow is much easier for a person to understand and a compiler to analyze.

5.4 Functions, Procedures, and Subroutines

Functions are sequences of instructions that, after being called, return to just after the point they were called. They are a powerful mechanism for structuring programs because they isolate parts of design and provide operations with well-defined inputs and outputs, the basic building block in complex systems.

The functions in high-level languages may be recursive in that a function may directly or indirectly call itself. Many problems are best solved using a divide-and-conquer approach that first splits a problem into two smaller subproblems. These subproblems are then split again and again until the problem is reduced to a trivial case. Recursive functions make it easy to describe such a solution. A recursive function generally begins by testing for a trivial problem. If the problem is not trivial, the function splits the problem and calls itself on the two or more parts.

A stack enables recursive functions. The stack stores return addresses and local copies of variables, one for each active invocation of the function. Calling a function pushes a return address on the stack as well as local storage; a return removes the local storage and pops the return address before jumping to it.

5.5 Interrupts

Because embedded systems often need to deal with real-time events, interrupts and interrupt service routines are common in embedded system software. An interrupt is a signal that tells a processor to pause what it is doing and deal with a real-time event. Once the interrupt is handled, the processor resumes what it was doing. A typical interrupt might be caused by new data arriving on a serial interface. An interrupt service routine would read this data from hardware and put it into a buffer in memory before returning.

Interrupt service routines are usually small and do the minimum amount necessary to avoid accidentally discarding data. Certain DSP s have especially efficient interrupt mechanisms that simply insert a single instruction into the stream when an interrupt occurs. This instruction generally copies a value from a location in memory (generally corresponding to memory-mapped I/O) into a a buffer and increments the buffer pointer. General-purpose processors usually have more complicated interrupt handling mechanisms that save processor state before running a service routine. In all cases, however, the main goal is to service an interrupt quickly before the condition that prompted it changes and is lost.

5.6   I/O

Processors are useless unless they are able to communicate. Certain processors have special instructions and interfaces just for I/O, but now most expect I/O to be memory mapped, i.e., to appear as normal memory locations. A serial interface is a typical example of a memory-mapped I/O device. One memory address might contain the latest byte read from the serial line, another, when written to, might send a byte, another location would set the bit rate for the interface, another might contain a status word with bits indicating whether a new word has been read, whether the last word was sent successfully, etc.

5.7   Memory Management

Managing memory in a software embedded system, like all software, can be a challenge. Embedded systems are often expected to run forever in restricted memory spaces, making it very important to avoid running out of memory. For this reason, embedded systems, like most operating systems, statically allocate all their memory at once, preferring fixed-sized buffers to variable-sized ones. This often places hard, sometimes arbitrary limits on the number, say, of active network connections, but this is often a reasonable price to pay for robustness.

5.8   ROM: Read-Only Memory

While most general-purpose computing systems store their programs on hard, fixed disks, most embedded systems, especially portable low-power ones, cannot afford the power consumption and weight of such disks and instead store their programs in read-only memory (ROM). This is semiconductor memory whose contents have been set to fixed values when it was manufactured. Like disks, the contents of ROM persist even when the power is turned off, but unlike disks, the contents may not be changed. Embedded system code must often be able to run from ROM, so it needs to be written so that sections that must be modified must be kept separate. Also, any initial values in modifiable memory must be explicitly initialized.

5.9   Device Drivers

A device driver is a piece of software responsible for interacting with a hardware peripheral such as a display or serial interface. It forms an abstraction layer between the low-level interface to hardware (typically setting and clearing bits in registers) and a higher-level interface that accepts commands such as "draw a rectangle" or "send a data packet." Creating device drivers is often tedious, requiring a good understanding of how the hardware functions as well as how the software will want to access it. This is especially difficult when hardware and software are developed in parallel since both specifications tend to change.

5.10   Exercises

5–1.   When would you use assembly language instead of a high-level language?

5–2.   When would you choose a floating-point number representation as opposed to a fixed-point? Which gives better accuracy for the same number of bits?

5-3.   What's the main advantage of a type system? What is the main disadvantage?

5-4.   Name five possible sources of interrupts in a typical workstation.

5-5.   Why is dynamically-allocated memory less appropriate in embedded systems than workstation applications?

5-6.   What key element do you need to implement recursive functions?

6

Assembly Languages

An assembly language program is a list of processor instructions written in a symbolic, human-readable form. Each instruction consists of an opcode and some operands in certain addressing modes. E.g., the instruction add r5,r2,r4 might add the contents of two registers and write the result to a third. Although most instructions are executed in order, branch instructions can send the processor's program counter (which points to the next instruction to execute) elsewhere to execute code, allowing for conditionals and loops.

Each processor has an assembly language characterized by its opcodes, addressing modes, registers, and memories. The opcode is the computation performed by an instruction (e.g., add two numbers, branch to a different instruction), and an addressing mode defines how and where the data for the computation is read and written (e.g., from a register, from a particular memory location, and from a memory location whose address is in a register). Memory is an uniform array of bytes used to store a program's data, and registers can be thought of as small, fast pieces of memory that can be accessed easily.

Unlike higher-level software languages, using an assembly language requires an understanding of the processor's architecture since the language manipulates it directly. While this chapter will discuss the architectures of a few processors, the field of processor architecture is far broader. An excellent starting point for further reading is the pair of books written by the two researchers responsible for the RISC revolution, Patterson and Hennessy [66, 37].

31	0	
	eax	Accumulator
	ebx	Pointer
	ecx	Loop, string counter
	edx	I/O pointer

15	0	
	cs	Code segment
	ds	Data segment
	ss	Stack segment
	es	Extra segment
	fs	Data segment
	gs	Data segment

	esi	Source index
	edi	Destination index
	ebp	Base pointer
	esp	Stack pointer

	eflags	Status word
	eip	Instruction pointer

Figure 6.1: i386 architecture registers. eax–edx are mostly general purpose, esi and edi are used for string instructions; esp and ebp are used for the stack and stack frames; the segment registers, now rarely used, are added to addresses to allow code and data to be separated.

6.1   CISC: The i386 Architecture

Intel's i386 architecture is a typical Complex Instruction Set Computer (CISC). This class is characterized by a small set of specialized registers (i.e., they cannot be used interchangably: Figure 6.1) used by a large, irregular instruction set whose instructions range from simple to very complex (Figure 6.4). CISC can be convenient for a human to program, especially when one of the more complicated instructions solves a particular problem, but is more difficult for a compiler.

The i386 architecture has four almost general-purpose 32-bit registers (eax–edx), two registers for string operations (esi, edi), a stack and base pointer (esp, ebp), a status word that holds flags such as carry and overflow, an instruction pointer (program counter), and six segment registers. The segment registers are a vestige of the architecture's sixteen-bit youth. To access more than 64K of memory, the code segment register was added to addresses during instruction fetch, another during stack operations, and a third during data operations. Thus code, data, and the stack could be placed in different 64k segments. Most software now runs with zeroed segment registers.

	Prefixes	Opcode	ModR/M	SIB	Disp	Imm
bytes:	0-4	1-2	0-1	0-1	0-4	0-4

Figure 6.2: i386 instruction format. Prefixes can influence string instruction behavior, the use of segment registers, and operand sizes. ModR/M and SIB specifies the addressing mode: type, registers, and a scale factor. The displacement is added to memory addresses. The final field is immediate data: a constant that can be, say, added.

```
    jmp   L2          # go to L2
L1:                   # label
    movl  %ebx, %eax  # n -> m
    movl  %ecx, %ebx  # r -> n
L2:
    xorl  %edx, %edx  # clear %edx
    divl  %ebx        # m / n
    movl  %edx, %ecx  # rem -> r
    testl %ecx, %ecx  # if r = 0,
    jne   L1          # go to L1
```

Figure 6.3: Euclid's algorithm in i386 assembly language. Symbols like %ebx represent registers. movl means "move long value." divl %ebx divides %eax by %ebx and puts the remainder in %edx. (Compare with Figure 5.1)

The i386 architecture presents a standard byte-addressed von Neumann memory model, i.e., both programs and data are stored in the same space. The i386 uses separate caches for programs and data to avoid this potential bottleneck, but this is largely transparent.

Like any byte-addressed processor, eight bits is the smallest unit that can be read or written, but the i386 also supports multi-byte data types. Unlike some processors, the i386 allows unaligned accesses, e.g., a four-byte word does not have to start at an address that is a multiple of four bytes. However, reading and writing unaligned objects is slow because it generally requires two memory operations.

The i386 has a special 64k-byte address space devoted to I/O. Special instructions (e.g., in, out) read and write to these addresses, which are assumed to be connected to peripherals such as serial ports. Although most systems now use memory-mapped I/O (where I/O devices appear as oddly-behaving memory locations) since it simplifies the instruction set, the I/O ports on a i386 have the benefit of behaving precisely as programmed: I/O instructions are executed in order and always wait before starting the next instruction.

Arithmetic

add adc sub sbb	Add, subtract	neg not	Negate, Not
mul imul	Multiply	div idiv	Divide
cmp test	Compare	inc dec	Inc/decrement
and or xor not	Logical	sal shl shld	Shift left
ror rol rcr rcl	Rotate	sar shr shrd	Shift right
bt bts btr btc	Bit test	bsf bsr	Bit scan
aaa aas aam aad	ASCII adjust	daa das	Decimal adjust
nop	No operation		

Data Transfer

mov	Move	movsx movzx	sign/zero extend
in out	Read/write I/O	xchg	Exchange
push	Push register	pop	Pop register
pusha pushad	Push registers	popa popad	Pop registers
lds les lfs lgs lss	Load far pointer	lea	Load address
cwd cdq cbw cwde	Convert	xlat xlatb	Table lookup

Flag Control

stc clc cmc	Carry	std cld	Direction
lahf sahf	Load/store	sti cli	Interrupt
pushf pushfd	Push	popf popfd	Pop

Jump and Branch

call ret	Call/return	enter leave	Call/return
int into bound	Call Interrupt	iret	Interrupt return
loop loopz loope loopnz loopne			Loop
jmp je jz jne jnz ja jnbe jae jnb jb jnae			
jbe jna jg jnle jge jnl jl jnge jle jng jc			Branch
jnc jo jno js jns jpo jnp jpe jp jcxz jecxz			

String Move/Compare

movs movsb movsw movsd	Move string
rep repe repz repne repnz	Repeat
cmps cmpsb cmpsw cmpsd scas scasb scasw scasd	Compare/scan
lods lodsb lodsw lodsd stos stosb stosw stosd	Load/store
ins insb insw insd outs outsb outsw outsd	Input/output

Conditional Set Byte

```
sete   setz   setne   setnz   seta   setnbe setae setnb
setnc  setb   setnae  setc    setbe  setna  setg  setnle
setge  setnl  setl    setnge  setle  setng  sets  setns
seto   setno  setpe   setp    setpo  setnp
```

Figure 6.4: Summary of i386 instructions

The i386 instruction set (Figure 6.4) is enormous and varied. Instructions range from a single byte (e.g., nop) up to a four-byte prefix followed by two bytes of opcode, two modifier bytes, four displacement bytes, and four immediate data bytes (Figure 6.2). This usually reduces code size, since simple, common instructions take less space, but presents a challenge to the processor's instruction decoder.

The add instruction illustrates the variety of addressing modes. The instruction adds the contents of its source (a register, memory location, or an immediate value) to its destination (a register or memory location) and stores the result in the destination, and can do this for a byte, word (16-bit), or doubleword (32-bit). Only one memory location can be used per instruction: both operands cannot come from memory. The memory location can be specified as the sum of a register, a register multiplied by 2, 4, or 8, and an immediate displacement of up to 32 bits.

```
addb $4, %al            # Add immediate 4 to byte register al
addl $999999999, %edx   # Add immediate doubleword to dx
addw -10(%ebp), %ebx    # Add word at bp-10 to ax
addb %dl, (%ebp,%esi)   # Add dl to memory at bp+si
addw %fs:-5(%ebx), %ecx # Add word at bx-5, accessed through
                        # segment register fs, to cx
addl -50(%ebp,%eax,4), %edx # Add word at bp + ax*4 - 50 to dx
addl $999999999, %es:88888888(%eax,%esi,2)
```

The first instruction is two bytes long (opcode plus single immediate byte), but the last instruction, because it uses a complex addressing mode and includes two four-byte constants, is twelve bytes long.

Many instructions greatly restrict the source of their operands. The div instruction, which divides unsigned values, is typical. The divisor can come from a register or from memory using a complex addressing mode, but the 64-bit dividend must be in edx:eax, and the quotient is placed in eax, the remainder in edx.

Other instructions perform very complex tasks. An extreme example is the enter instruction, which creates a stack frame for a newly-called procedure. It takes two parameters: the amount of storage to reserve on the stack for automatic variables and the lexical nesting depth of the procedure, used in a language such as Pascal that permits a function defined within another to access the outer function's variables. It pushes ebp on the stack, then repeatedly pushes the data in memory at ebp onto the stack, and finally adjusts the stack to make room for storage. In pseudo code, this is:

```
push ebp
framePtr = esp
if (level > 0) {
  repeat level-1 times # Create one display per nesting level
    ebp = ebp - 4
    push (ebp)        # Push data in memory at ebp
  push framePtr
ebp = framePtr
esp = esp - size
```

The i386 instruction set includes many control-flow instructions. The simplest is jmp, which always sends control to a different location. Even jmp comes in many flavors: relative with an eight-, sixteen-, or thirty-two-bit offset, useful for branching within a small segment of code, or absolute with a sixteen- or thirty-two-bit immediate or in-memory address.

Conditional branches send control to another instruction depending on one of thirty-two different conditions, typically set by the cmp instruction, which subtracts source from destination, sets condition flags, and discards the numeric result.

```
cmpl %eax, %ebx
jg L1             # Branch if ebx > eax (signed)
ja L2             # Branch if ebx > eax (unsigned)
jle L3            # Branch if ebx <= eax (signed)
je L4             # Branch if ebx = eax
jcxz L5           # Branch if cx = 0
jecxz L6          # Branch if ecx = 0
test %eax, %eax
jpe L7            # Branch if eax has even parity
```

6.2   RISC: The MIPS Architecture

The MIPS architecture is a typical example of a so-called Reduced Instruction Set Computer (RISC). This class is characterized by a highly orthogonal, simplified instruction set (Figure 6.8) centered around a large group of general-purpose registers (Figure 6.5). Arithmetic operations operate exclusively on registers and memory accesses only copy data to and from the register file. Such a load/store architecture is simpler because it eliminates most memory access exceptions and hence the need to resume most instructions.

The MIPS architecture has thirty-two 32-bit general-purpose registers, a 32-bit program counter, and a pair of 32-bit registers to store the results of multiplication or division (Figure 6.5). Two of the thirty-two registers are special: $0 always reads zero and discards values written to it, the jump-and-link instruction stores its return address in $31.

The instruction set is very regular and orthogonal, making it easier to implement and write compilers for. Every instruction is thirty-two bits and is encoded in one of three main ways (Figure 6.8). This allows for relatively few addressing modes:

```
add  $1,$2,$3  # Register: add $2 and $3, store result in $1
add  $1,$2,10  # Immediate: add $2 and 10, store result in $1
lw   $1,10($2) # Offset: Read memory at $2 + 10, store in $1
beq  $1,$2,100 # PC relative: Branch to PC + 100 if $1 = $2
j    1000      # Immediate long: Jump to address 1000
```

The three-register arithmetic instructions are very flexible and can be used in surprising ways. For example, the MIPS has no explicit register-to-register move instruction. Instead, an assembler interprets something like move $1, $2 as addu $1, $2, $0, i.e., adding the constant zero in $0 to $2 and storing it in $1. While this may seem wasteful, adding an explicit move instruction would just complicate the processor implementation and run no faster, since an add is one of the fastest instructions.

Many processors include multi-word instructions that can, e.g., load large constants into a register. The MIPS instead supplies the lui instruction that loads a sixteen-bit constant into the upper halfword of a register. An immediate or can then set the lower sixteen bits. Although this takes two cycles, an explicit immediate load instruction would take as long, and since most constants are small, the lui is rarely used anyway.

The MIPS jump and branch instructions have a delay slot: the instruction immediately following a jump or branch is executed regardless of whether the branch is taken. Exposing the processor's pipeline simplifies executing programs quickly. Modern processors are usually

31   General Purpose   0
$0 = 0
$1
⋮
$30
$31 (link)

31   Multiply/Divide   0
HI
LO

31   Program Counter   0
PC

Figure 6.5: MIPS architecture registers. Register $0 always returns 0; the jal instruction stores a return address in $31. Multiplication and division store their results in HI and LO.

pipelined for speed, meaning each instruction is broken into phases such as fetch, decode, read, compute, and write. At any time, a few instructions are running concurrently, each at a different phase. The result is that when a branch instruction has been decoded, the next instruction has already been fetched. Rather than always discard this instruction, the MIPS executes it. A clever programmer or compiler can put a useful instruction in that slot, such as the move instruction immediately following the final j instruction in Figure 6.7.

One philosophy behind the MIPS architecture is to keep the instructions as basic as possible to simplify the task of a compiler, which prefers instructions with few side-effects. An example is the jal instruction, intended to be a function call, which saves the address of the second instruction following it (i.e., after the delay slot) in $31. To return, a function need only jump to the contents of register $31 (the second-to-last instruction in Figure 6.7). Other processors have complex instructions that remove a return address from a stack, adjust stack and perhaps frame pointers, and various other tasks. The MIPS shines in cases where all these extra operations are not needed, since it is not forced to do them.

6.3   Harvard Architecture DSP s: The 56000 Architecture

RISC and CISC machines are designed for general-purpose computing tasks. They handle control- and arithmetic-intensive applications with equal aplomb, but can be needlessly slow, power-hungry, and expensive for an embedded system designed for a single task.

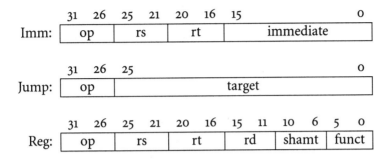

Figure 6.6: MIPS instruction encoding. All instructions are 32 bits. The op + funct fields contain the opcode, immediate and target fields hold constant addresss, offsets, and data, shamt is a shift amount for shift instructions, rs, rt, and rd define source, target, and destination registers.

```
gcd:
  remu $4,$4,$5
  beq  $4,$0,L4
L6:
  move $2,$5      # m <- n
  move $5,$4      # n <- r
  remu $4,$2,$5 # r <- m mod n
  bne  $4,$0,L6 # r=0, go to L6
L4:
  j    $31
  move $2,$5
```

Figure 6.7: Euclid's algorithm as a MIPS assembly language function. remu is a macro that expands to divu, bnez, nop, break, and mfhi. move is interpreted as an add with zero.

Arithmetic

nop	Do nothing		
add	Add	addu	Add unsigned
addi	Add immediate	addiu	Add immediate unsigned
sub	Subtract	subu	Subtract unsigned
subi	Subtract immediate	subiu	Subtract immediate unsigned
mult	Multiply	multu	Multiply unsigned
div	Divide	divu	Divide unsigned

Logical

and	Logical AND	andi	Logical AND immediate
or	Logical OR	ori	Logical OR immediate
xor	Logical XOR	xori	Logical XOR immediate
sll	Shift left logical	sllv	Shift left logical variable
srl	Shift right logical	srlv	Shift right logical variable
sra	Shift right arithmetic	srav	Shift right arithmetic variable

Load/Store

lw	Load word		
lh	Load halfword	lhu	Load halfword unsigned
lb	Load byte	lbu	Load byte unsigned
lui	Load upper immediate		
sw	Store word	sh	Store halfword
swl	Store word left	swr	Store word right
sb	Store byte		
mfhi	Move from HI	mflo	Move from LO
mthi	Move to HI	mtlo	Move to LO

Conditional

slt	Set on less than	sltu	Set on less than unsigned
slti	immediate	sltiu	immediate unsigned

Jump and Branch

j	Jump	jr	Jump register
jal	Jump and link	jalr	Jump and link register
beq	Branch on equal	bne	Branch on not equal
bltz	less than zero	bgtz	greater than zero
blez	less than or equal to zero	bgez	greater than or equal to zero

Figure 6.8: Summary of MIPS instructions

Digital signal processing, which transforms high-speed streams of data representing continuous signals such as voice or data, is a common application of this type. Real-time signal processing usually has strict timing requirements, but the algorithms are usually very predictable. Arithmetic is their main requirement; they rarely use data structures more complex than circular buffers and their control-flow usually consists of little more than counted loops.

Digital signal processors or DSP s such as Motorola's 56000 are optimized to execute these algorithms with minimal overhead. The 56000 is a typical Harvard Architecture DSP, posessing an ALU optimized for the multiply-and-accumulate (MAC) operation, a small set of special-purpose registers (even more specialized than those in a CISC machine), and separate program and data memory spaces (this separation gives it the Harvard designation; processors with a single memory are termed von Neumann).

Like most DSP s, the 56000 is designed around an efficient (single-cycle) MAC operation, which multiplies two numbers and accumulates (adds) the result in a register. The finite impulse response (FIR) filter algorithm, a common application, uses a MAC in its inner loop. For each output sample, an FIR filter computes the dot product of a vector of the past n input samples $x_t, \ldots, x_{t-(n-1)}$ and a vector of n filter coefficients $a_0, \ldots, a_{n-1}$, i.e., $\sum_{i=0}^{n-1} a_i x_{t-i}$.

Dsp s eschew the large, general-purpose register files of RISC machines in favor of small, specialized register files for reasons of power, speed, and area. The 56000 (Figure 6.9) has a group of registers for the datapath, another group for the address generation unit, and a third for program control. Transferring data between registers in different groups is possible, but slow.

Designed around an efficient single-cycle MAC operation, the datapath (Figure 6.11) sports four 24-bit source registers used as ALU operands and two accumulators that store ALU results. Every cycle, the datapath can read two operands into the X and Y registers from memory, multiply the previous contents of these two registers, and add the product to an accumulator.

The address generation unit was mainly designed to implement circular buffers. An address usually comes directly from one of the Pointer registers, which as a side-effect may also have its correspond-

	55 48	47 24	23 0
Source		X_1	X_0
Registers		Y_1	Y_0

	55 48	47 24	23 0
Accumulator	A_2	A_1	A_0
Registers	B_2	B_1	B_0

	Pointer	Offset	Modifier
	15 0	15 0	15 0
	R_7	N_7	M_7
	R_6	N_6	M_6
Address	R_5	N_5	M_5
Registers	R_4	N_4	M_4
	R_3	N_3	M_3
	R_2	N_2	M_2
	R_1	N_1	M_1
	R_0	N_0	M_0

15 0	15 0	
Loop Address	Loop Count	

	15 0	15 0	
Program Control Registers	Program Counter	Status Register	6 0
	PC Stack 15	SR Stack 15	Stack Ptr.
	⋮	⋮	
	0	0	

Figure 6.9: 56000 registers. Source registers feed operands to the adder and multiplier, accumulators hold the results. Pointer registers store memory addresses; offset registers define the stride for auto add/subtract operations; modifier registers select among flat, modulo, and bit-reversed addressing. The loop address register holds the address of the end of a loop; the loop counter holds the count of that loop. The program counter holds the address of the currently-executing instruction. The stack is a pair of sixteen-bit-wide register files: one holds the old program counter, the other the old status register. The stack pointer includes status bits.

Arithmetic

mac	Multiply-accumulate	macr	Multiply-accumulate, round
adc	Add with carry	add	Addition
sbc	Subtract with carry	sub	Subtract
mpy	Multiply	mpyr	Multiply and round
div	Divide iteration		
asl	Arithmetic shift left	asr	Arithmetic shift right
addl	Shift left and add	addr	Shift right and add
subl	Shift left and subtract	subr	Shift right and subtract
norm	Normalize	rnd	Round
abs	Absolute value	neg	Negate
clr	Clear a registero		
tst	Test operand		
cmp	Compare	cmpm	Compare Magnitude
t..	Transfer conditionally	tfr	Transfer register

Logical

and	Logical AND	or	Logical OR
andi	AND immediate (control)	ori	OR immediate (control)
eor	Logical XOR	not	Logical complement
lsl	Logical shift left	lsr	Logical shift right
rol	Rotate left	ror	Rotate right
bclr	Bit test and clear	bset	Bit test and set
bchg	Bit test and change	btst	Bit test memory/reg.

Move

lua	Load updated address		
move	Data register	movec	Control register
movem	Program memory	movep	Peripheral data

Program Control

do	Start loop	enddo	Exit loop
rep	Repeat next instruction		
j..	Conditional jump	jmp	Jump
jclr	Jump if bit clear	jset	Jump if bit set
js..	Conditional call	jsr	Jump subroutine
jsclr	Call if bit clear	jsset	Call if bit set
nop	No operation	reset	Reset peripherals
rti	Return from interrupt	rts	Return from subroutine
stop	Stop processor		
swi	Generate interrupt	wait	Wait for interrupt
il	Illegal instruction		

Figure 6.10: Summary of 56000 instructions

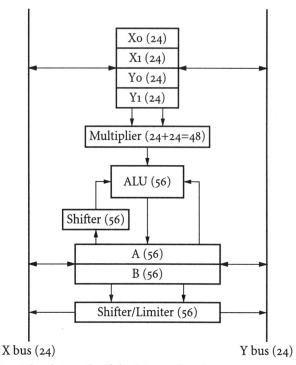

Figure 6.11: The datapath of the Motorola 56001 DSP

```
n       equ 20     ; Number of taps (symbolic constant)
wddr    equ $0     ; Start of sample data (X memory)
cddr    equ $0     ; Start of coefficient data (Y memory)
input   equ $ffe0  ; I/O locations
output  equ $ffe1

  move #wddr,r0  ; r0 is start of samples
  move #cddr,r4  ; r4 is start of coefficients
  move #n-1,m0   ; Set sample buffer size to 20
  move m0,m4     ; Set coefficient buffer size to 20

  movep y:input,x:(r0)                          ; read sample
  clr   a         x:(r0)+,x0  y:(r4)+,y0        ; clear and read first
  rep   #n-1                                     ; repeat 19 times
  mac   x0,y0,a   x:(r0)+,x0  y:(r4)+,y0        ; a += x0*y0, fetch next
  macr  x0,x0,a   (r0)-                          ; multiply last
  movep a,y:output                               ; output filtered sample
```

Figure 6.12: A 20-tap FIR filter in 56000 assembly language

ing Offset register added or subtracted modulo the corresponding Modifier register. Thus a circular buffer is defined by placing its base address in a pointer register, its size in the matching modifier register, and a stride in the matching offset register.

The 56000 has hardware support for the bit-reversed addressing used in the Fast Fourier Transform, another common signal processing algorithm. This reverses the direction of the carry propagation in address arithmetic when a modifier register is set to 0.

The 56000 has three memory spaces: program, X, and Y. The X and Y spaces are intended for data, and they are split so that sample data can be put in one and coefficients in another. Since each is capable of a single read or write each cycle, two data memories makes it possible to supply operands to the MAC once per cycle.

Instructions on the 56000 are twenty-four bits wide and provide restricted parallelism. Arithmetic instructions, such as mac, allow the most. A single mac can multiply an X and Y register, accumulate the result in one of the accumulators, and simultaneously load the same X and Y registers from the X and Y memories and update (increment) address registers. Only a few instructions allow this much parallelism and those that do only allow a narrow range of parallel operations.

Dsp algorithms often use simple counted loops where the number of times the loop should run is a simple fixed number, so DSP s often provide direct support for loops in the instruction set. The 56000 can repeat a single instruction (typically a MAC) a fixed number of times with no delay between invocations. The decoded instruction is simply held in the processor's instruction register and executed repeatedly.

The FIR filter code in Figure 6.12 illustrates the 56000's strengths. It begin with a series of equates that define symbolic constants, e.g.,

```
n equ 20
```

defines n, the number of taps, to be 20. Next are a series of moves that set up two pairs of pointer and modifier registers. r4 was chosen because it will be used to generate addresses for Y memory.

The clr instruction,

```
clr    a        x:(r0)+,x0   y:(r4)+,y0
```

illustrates the typical form of an arithmetic instruction: the opcode and first group of operands control the datapath, and the second and third groups control the address generation unit and X and Y data busses. Here, it clears accumulator a, reads the first sample from X memory at the address in r0, stores it in x0, increments r0 (since m0 is set to 19, wrapping around to 0 if r0 has reached 20), and does the same to read the first coefficient at r4 in Y memory.

The rep #n-1 instruction repeats the following mac instruction 19 times. In each cycle, the mac multiplies the x0 and y0 registers (whose values came from the preceding clr or the last mac) and adds the result to the accumulator a. The mac address/bus behavior is the same as the clr: it reads the next sample and coefficient and updates the address registers.

Finally, after the 19th mac, the last coefficient/sample pair is in the x0 and y0 registers, and the macr multiplies them, accumulates and finally rounds the result. At the same time, the r0 register is decremented to prepare room for the next sample.

This example illustrates how the effect of the pipeline is felt in 56000 assembly language. Although the mac instruction reads a sample and coefficient from memory, it actually multiplies the last pair that was read. Hence, a loop prologue (here, the memory operations on the clr instruction) and an epilogue (the macr) is needed to handle the end cases.

6.4   VLIW DSP s: The 320c6 Architecture

Very long instruction word (vliw) processors consist of eight or more instruction units operating in unison. They are controlled by a single instruction stream (i.e., a single program counter), and each instruction can control each unit independently, requiring very long instruction words, hence the name. The instruction set of a vliw resembles that of a risc processor: instructions are simple, orthogonal, and access a large, general-purpose register file. But a vliw also provides the explicit instruction-level parallelism typical of a dsp.

The additional flexibility and larger number of heavily-pipelined instruction units makes a vliw very difficult to program by hand, but compiler technology has evolved to where efficient programs can be compiled from C code.

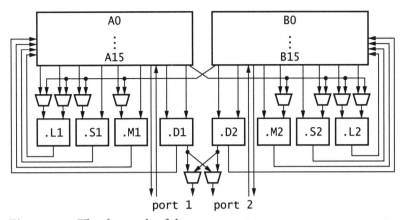

Figure 6.13: The datapath of the TMS320C6201 VLIW DSP

Texas Instruments' TMS320C6201 and TMS320C6701 are a pair of commercially successful VLIW processors targeted at DSP applications. The fixed-point 'c62 runs at 200 MHz; the floating-point 'c67 runs at 167 MHz. Both are 32-bit processors; the 'c67 also supports 32- and 64-bit IEEE 754 floating-point.

Each processor's datapath is divided into two identical halves (Figure 6.13), each with four functional units and a bank of sixteen 32-bit general-purpose registers that can also be treated as eight 64-bit registers. There are four types of functional unit: logical/arithmetic (.L), shifter/branching (.S), multiplication (.M), and data (memory) operations (.D). Most instructions are specific to a particular functional unit, but certain simple instructions, such as integer addition, can be performed on all but the multiplier units.

The processors also have special-purpose registers that control behavior that rarely needs to change. An addressing mode register affects the generation of addresses for circular buffers. Another register controls interrupt behavior. A register on the 'c67 controls rounding and other floating-point behavior.

These processors use a Harvard architecture. Data memory is separate from program memory. Program memory is fetched 256 bits at a time, allowing a full eight instructions to be fetched at once.

Each instruction issue consists of between one and eight actual instructions (the rest are implicitly NOPs). Each instruction for a function unit is 32 bits long, and each uses one bit to control parallel instruction issue. If this least significant "p" bit in an instruction is set, the next instruction in sequence will be executed in parallel with it. Nop instructions on these processors have an optional count that allows them to delay multiple cycles to flush the pipeline.

Both processors are heavily pipelined. Instruction fetch takes four cycles, decode takes two, and execution takes anywhere from one to nine cycles, depending on the complexity of the instruction. That it takes four instructions to fetch an instruction reflects how memory is slower than registers. The four cycles generate the address, send it, wait, and finally receive.

The deep instruction fetch pipeline delays the effect of a branch. It takes five cycles for a branch to have an effect: one to actually execute the branch and four to fetch the branch target instruction. The decision to branch, therefore, is best made long before it is necessary to branch. It is often possible to rearrange instructions so that a branch comes earlier, but this can be difficult.

The processors disable interrupts when a branch is in progress because the deep pipeline makes it difficult to resume halfway through a branch instruction.

Because of the extreme level of parallelism and branch latency, the architectures provide conditional instructions that are only executed if a particular flag in a particular condition register is set. Instead of loops filled with conditional branches, this mechanism allows for loops consisting of clean, linear sequences of conditional instructions. Also, conditional instructions do not have the latency problems of branches because they are fetched and decoded regardless of whether they execute.

Instruction decode takes two cycles: one to distribute the parallel instructions to the functional units based on their "p" bits, the other to perform traditional instruction decoding: preparing the ALU, decoding register selects, and so forth.

The actual execution of an instruction can take one cycle for simple arithmetic operations such as addition, and as much as nine cycles for multiplication and floating-point operations on the 'c67. It is interesting to note that getting an instruction ready to execute can take six times as long as actually executing it on this architecture. Reading data from memory is slow (five cycles) because the instruction must produce the address and wait for the data to come back. Storing, by contrast, only takes three cycles because there is no need to wait for data to return.

Both processors expose the pipeline in the sense that an assembly language programmer must understand what data is produced when. Also, many pairs of instructions cannot be issued in parallel because they would use the same resource. The processors have locks that keep track of when certain instructions cannot be issued and simply discard offending instructions to prevent damage to the hardware.

Because of a limited number of register ports, memory paths, and other resources, there are many restrictions on what instructions can run in parallel. For example, each half of the datapath can only read one register from the other datapath in each cycle. Each datapath can only compute one memory address per cycle. Only one 64-bit value can be written per datapath per cycle. Each register can be read no more than four times each cycle, and written at most once.

Texas Instruments has produced an excellent C compiler for these architectures. The datapaths of these processors are complicated, but are complicated in a fairly regular way. Registers are general-purpose, and although communication between the two halves of the datapath is restricted, it is restricted in a uniform way. All of these factors (plus the fact that the architecture and compiler for it were developed concurrently) combine to make it possible for very sophisticated compiler technology to produce very good code. Naturally, this is a very good thing since the datapaths are complex enough to make it difficult for a human to simultaneously satisfy all constraints and produce efficient code.

Here is a small but useful 'c62 program fragment that implements an FIR filter. This is a single instruction word that uses six of the eight functional units to calculate $\sum_{i=1}^{n} a_i b_i$, illustrating how much ILP is possible:

```
FIRLOOP:
         LDH   .D1   *A1++, A2     ; Read coefficient
   ||    LDH   .D2   *B1++, B2     ; Read data
   || [B0] SUB  .L2   B0, 1, B0    ; Decrement counter
   || [B0] B    .S2   FIRLOOP      ; Branch if not zero
   ||    MPY   .M1X  A2, B2, A3    ; Form product
   ||    ADD   .L1   A4, A3, A4    ; Accumulate result
```

The double vertical bars (||) indicate all six of these instructions are issued in parallel. The two LDH instructions use the two data units (.D1 and .D2) to load 16-bit half words from the addresses in registers A1 and B1 into registers A2 and B2. The two address registers are also incremented (indicated by ++). The notation [B0] indicates the SUB and B instructions are conditional and execute only if the B0 register is non-zero. The SUB instruction uses the second logical unit to decrement the loop index in register B0, and the B instruction branches to the beginning of the loop.

The MPY instruction does the main work of multiplying the two coefficients (in registers A2 and B2) and placing the result in A3. The MPY is assigned to multiplication unit .M1, uses a cross path to fetch data from register B2 as indicated by the X. The final instruction accumulates the result by adding the result of the multiply in A3 to the result accumulating in A4.

This example does not show the loop prologue and epilogue—the instructions that must come before and after the loop to prime and later flush the pipeline. For example, the latency of the multiply instruction means the accumulation instruction actually sees the multiplication from a few clock cycles previous. Before this loop is entered, there must be enough results (or zeros) in the pipeline for the accumulation.

6.5   Exercises

6–1.   How would you perform bit-reversed addressing for an FFT if you didn't have special hardware to do it? What effect would this have on the speed and size of the code?

6–2.   Name two reasons RISC machines have largely replaced CISC machines. Name two reasons why you might still prefer a CISC machine.

6–3.   Why are traditional DSPs difficult targets for high-level languages? Why did their designers choose to make them that way?

6–4.   Name the main advantage and disadvantage of assembly language that exposes a pipeline.

6–5.   Why do the 'c62 and 'c67 processors disable interrupts during a branch?

6–6.   Write a very small (say, five or ten lines) C program and generate assembly code for it by giving the C compiler the -S option (on most, this generates a file named "foo.s" from a file "foo.c" that contains assembly language). Compare the number of instructions produced with and without the -O option (optimization). List a few of the optimizations the -O option performs. What effect does this have on the size of the generated code? The speed?

7

The C Language

The C language is the most successful software language ever devised. Developed between 1969 and 1973 along with the Unix operating system, it has always been a pragmatic language for solving its developer's problems. Limitations of early hardware (it first ran on a DEC PDP-11 with 24K of memory) forced it to be a small language, but this is one of its strengths. Since C's simple semantics closely fit those of typical processors, C compilers produce compact, efficient code.

A C program consists of functions built from sequences of arithmetic expressions structured with loops and conditionals. The instructions in a C program run sequentially: expressions are evaluated one after another. Control-flow constructs can send this single point of control back to the beginning of a loop or down a single branch of a conditional. A function call is more complex: when one is encountered (in an expression), control is passed to the called function, which runs until it produces a result, and control returns to continue evaluating the expression that contained the function call.

C programs use three types of memory. Space for global data is allocated when the program is compiled, the stack stores automatic variables allocated and released when their function is called and returns, and the heap supplies arbitrarily-sized regions of memory that can be deallocated in any order.

C provides pointers, a type of variable that may refer to data in any memory region. Pointers are a powerful abstraction that can be difficult to analyze because they introduce aliases for objects.

Hundreds of books have been written on C, but Brian Kernighan and Dennis Ritchie (the designer of the language) wrote what most consider the best [49]. That this slim (272-page) volume can convey most of what you need to know to program in C is a testament to the simplicity and power of the language. Serious programmers or implementers will want a copy of the reference manual [34]. Plauger [68] describes the standard C library and details an implementation.

7.1 Overview

A C program (e.g., Figure 7.1) is a collection of functions, global variables, and data types. All the action takes place in the functions, which are sequences of instructions with their own local storage. Control passes to a function when it is called, i.e., referred to in a statement being executed. A call passes control the function, which runs until it returns and passes control back to where it was called.

Each function consists of statements that are executed one at a time. When a statement receives control, it usually reads and writes information in memory before passing control to the next statement. These semantics are the same as most assembly languages; C is in some sense a high-level assembly language.

Statements come in two flavors. Expressions are an imperative version of their mathematical counterparts: they read and combine values from variables and functions to produce a result that is usually written to another variable. The other type of statement affects which statements are executed. Examples include loops, which execute a group of statements repeatedly, and conditionals, which execute different statements depending on the value of an expression.

The language also provides types that define how to interpret data in memory; variables that attach symbolic names to data; a standard library of functions that provides string manipulation, memory management, I/O, math functions, and other facilities; and a preprocessor for defining symbolic constants and macros, conditional compilation, and sharing information among multiple source files.

The language does not provide direct support for I/O (the standard library does), concurrency (an operating system can), or automatic garbage collection.

```c
#include <stdio.h>
#define MAXLINE 1024
char line[MAXLINE];

/* True if string matches pattern */
int contains(const char *string, const char *pattern)
{
  const char *s, *p;
  for (; *string != 0 ; ++string) {
    for (s=string, p=pattern ; *p != 0 ; ++s, ++p)
      if (*s != *p) break;
    if (*p == 0) return 1; /* Matched pattern */
  }
  return 0;
}

void printmatching(const char *pattern, FILE *fp)
{
  while (fgets(line, MAXLINE, fp) != 0)
    if (contains(line, pattern)) printf("%s", line);
}

int main(int argc, char *argv[])
{
  const char *pattern;
  FILE *fp;
  char *prg = argv[0];  /* Save program name */
  argc--, argv++;  /* Advance over over program name */
  if (argc < 1) {
    fprintf(stderr, "Usage: %s pattern [files...]\n", prg);
    return 1;
  }
  pattern = argv[0];
  if (argc == 1) printmatching(pattern, stdin);
  else
    while (--argc > 0)
      if ((fp = fopen(*++argv, "r")) == NULL) {
        fprintf(stderr, "%s: could not open %s\n", prg, argv[0]);
        return 1;
      } else {
        printmatching(pattern, fp);
        fclose(fp);
      }
  return 0;
}
```

Figure 7.1: A C program that prints lines that match a pattern

7.2   Types

Ultimately, C programs manipulate four basic datatypes, char, int, float, and double, which generally match types directly supported by the target processor. This makes for efficient code, but can mean C programs behave differently when compiled on different processors.

The most commonly-used basic type, int, represents a signed integer. The char type represents a single character, usually using eight bits to represent an ASCII character. The float and double types represent single- and double-precision floating-point numbers.

Types can be built into arrays (groups of objects of the same type that can be indexed by integers), pointers (references to objects in memory), structures (aggregations of types with named fields), and unions (objects that can take one of a few different types).

A C declaration associates a name with a type. The syntax is a series of type specifiers followed by a declarator. The specifiers describe a basic type or a simple variant of one (e.g., unsigned int). A declarator is a name possibly surrounded by nested syntax that indicate pointers, arrays, and functions. If the declarator appears in an expression, it produces the basic type described in the specifier. E.g.,

```
int i;                      /* i: an integer */
int *j, k;                  /* j: a pointer to an integer
                               k: an integer */
unsigned char *ch;          /* ch: pointer to a nonnegative
                               character */
float f[10];                /* f: array of ten floats */
char nextChar(int, char*);  /* nextChar: two-argument function
                               returning a char */
int a[3][5][10];            /* a: array of 3 arrays of 5 arrays
                               of 10 ints */
int *func1(float);          /* func1: one-argument function
                               returning pointer to integer */
int (*func2)(void);         /* func2: pointer to zero-argument
                               function returning integer */
```

New types can be named with typedef. For example, func2 might be more clearly defined

```
typedef int func2t(void); /* func2t is now a type */
func2t *func2;
```

A C structure is an object with named fields. For example,

```
struct {
  char *name;
  int x, y;
  int h, w;
} box;
```

defines the variable box to be an object of a structure type consisting of a character pointer and four integers. The dot operator provides access to the members of this structure, e.g., box.name, box.x.

A name following struct introduces a name for the struct, which is useful for avoiding repetition (although using typedef is more common) and for building self-referential structures:

```
struct tree {
  int val;
  struct tree *leftChild, *rightChild;
};

struct foo { int x, y; };
struct foo bar1;
typedef struct { int x, y; } foo_t;
foo_t bar2;    /* Equivalent to bar1 */
```

C provides a way to tightly pack data in structures. The size of integer-valued members may be specified, and the compiler will try to pack them together into words. For example, a compiler, given the structure

```
struct bits {
  unsigned int first_bit : 1;
  unsigned int two_bits : 2;
  unsigned int flag : 1;
};
```

might pack the three fields into a byte or word. Packed structures always work within a program, but since the compiler is free to choose the order in which fields appear within words, using them to pick apart bits stored in a file generally does not work. Furthermore, using bit fields generally makes for larger, slower code because of the need for shift and mask instructions for each read and write.

Unions provide another way to conserve space. Like structures, they contain named members of different types, but they only store one of these members at any given time. The programmer is responsible for always reading the last-written member; violating this usually causes a garbage value to be read. Unions have the same syntax and usage as structures, but use the union keyword.

```
union one_of {
  int i;
  float f;
  double d;
};
```

Unions allow arrays to contain objects of different types and provide a limited form of polymorphism. For example, they can be used to build a hashtable that stores integers and strings.

7.3   Variables and Storage

C programs can use three types of storage. Global or static storage, allocated when the program is compiled, is best for fixed-size data needed during a complete run of the program. Automatic storage, allocated and released when a function is called and returns, is for data only needed by a particular invocation of a function. Finally, the standard library provides routines that allocate and release arbitrarily-sized blocks from the heap. The heap is for data whose size or quantity is not known until the program runs.

Here is how storage may be allocated:

```
#include <stdlib.h>

int global_static;
static int file_static;

void foo(int auto_param)
{
  static int func_static;
  int auto_i, auto_a[10];

  int *auto_v = alloca(sizeof(int) * auto_param);
  double *auto_d = malloc(sizeof(double) * 5);
}
```

Space for the three integers `global_static`, `file_static`, and `func_static` is allocated when the program is compiled and remains fixed while the program executes. The `global_static` variable can be made visible to all files that comprise the program. Other files that wish to access this variable must declare

```
extern int global_static;
```

The variable `file_static` is only visible to functions in the same file; others may not declare it `extern`. Similarly, `func_static` is only visible within the function. A single incarnation of it always exists.

Space for the variables `auto_i`, `auto_a`, `auto_v`, and `auto_d` is allocated on the stack when the function is entered and released when it returns. If a function calls itself, each call has a separate copy of these variables.

The pointer `auto_v` is initialized with a call to `alloca`, which returns a pointer to an arbitrarily-large block of space on the stack that will be freed when the function returns. The `alloca` function is not part of the standard, but included in many implementations, this is a way to work around C's restriction of constant array bounds for automatic variables. (C prohibits a declaration like `int a[len]` if `len` is a variable or function parameter.)

The pointer `auto_d` uses the standard C library's `malloc` function to request an area of memory that will persist until it is freed with a call to `free(auto_d)`.

Dynamic memory management is a powerful but dangerous mechanism. Some of the most common and insidious bugs in C programs are due to accessing pointers whose space has been freed or not yet allocated. Since freed data is rarely overwritten immediately, the program will work unpredictably.

Memory leaks are another problem. These occur when the location of allocated memory is lost. For example, the following is legal:

```
char *c = malloc(sizeof(char) * 10240);
c = malloc(sizeof(char) * 10);
/* Location of the first 10K region is now lost */
```

Memory leaks rarely cause terminating programs to fail, but they are serious in embedded systems where programs must run indefinitely.

Managing dynamic memory is slower and more wasteful than using either static or automatic storage. Allocating and releasing memory in an arbitrary order demands an unpredictable, time-consuming algorithm. Each allocated block generally consumes two additional words in memory (the size of the block and a pointer to the next). Knuth [50] provides a good introduction to algorithms for managing dynamic memory.

Despite all these problems, most large C programs use dynamic memory. Systems with strict real-time performance requirements are often willing to accept fixed limits (e.g., a maximum number of running processes) in return or predictibility. The C++ language provides a higher-level mechanism that calls `malloc` and `free`; Java's automatic garbage collection frees memory automatically.

7.4   Expressions

Expressions perform all the useful work in a C program. They read and write memory, perform arithmetic, and call functions. They are built from numbers, variables, function calls, and operators to combine these. C provides a wide variety of operators, listed in Figure 7.2.

Pointers are important in C, and many operators manipulate them. The simplest is *, which dereferences a pointer. If p is a pointer, then *p is what it points to. The -> operator is a shorthand for dereferencing a pointer to a structure and choosing a field: (*p).m can be written more clearly as p->m.

C allows simple pointer arithmetic. If p is a pointer into an array of integers, then p+3 refers to the third element after p. If p is the base of the array, this is equivalent to p[3]. If p and q are pointers into the same array, then p-q is the number of elements between them. Arithmetic between pointers in different arrays is undefined.

C's bitwise operators support bit manipulation with logical operators AND, OR, exclusive-OR, and NOT, and left and right shifting operators. The shift operators fill with zeros or ones depending on whether the value is signed or unsigned.

An increment/decrement operator modifies its argument, usually a single variable. The prefix variant returns the value after it is changed; a postfix operator returns the value before changing it.

`f(r,r,...)`	function call	`a[i]`	array index
`p->m`	means `(*p).m`	`s.m`	struct member
`!b`	logical NOT	`~i`	bitwise NOT
`-i`	negation		
`++l`	preincrement	`--l`	predecrement
`l++`	postincrement	`l--`	postdecrement
`*p`	follow pointer	`&l`	address of
`(type) r`	typecast	`sizeof(t)`	size of a type
`n * o`	multiplication	`n / o`	division
`i % j`	remainder		
`n + o`	addition	`n - o`	subtraction
`i << j`	shift left	`i >> j`	shift right
`n < o`	less than	`n > o`	greater than
`n <= o`	... or equal	`n >= o`	... or equal
`r == r`	equal	`r != r`	not equal
`i & j`	bitwise AND		
`i ^ j`	bitwise XOR		
`i | j`	bitwise OR		
`b && c`	lazy logical AND		
`b || c`	lazy logical OR		
`b ? r : r`	conditional		
`l = r`	assignment		
`l += n`	add assign	`l -= n`	subtract assign
`l *= n`	multiply assign	`l /= n`	divide assign
`l %= i`	remainder assign	`l &= i`	AND assign
`l ^= i`	XOR assign	`l |= i`	OR assign
`l <<= i`	shift left assign	`l >>= i`	shift right assign
`r1 , r2`	sequence		

Figure 7.2: C's operators, grouped by precedence (top is highest). a is an array, b and c are boolean, f is a function, i and j are integers, l is an lvalue, m is a structure or union member name, n and o are numbers, p is a pointer, r is an rvalue, t is a type.

The lazy logical operators && and || only evaluate their right arguments if necessary: if the left argument of && evaluates false, the expression is false and the right argument is not evaluated. Such short-circuit evaluation is usually faster with these semantics because it can escape early. This also avoids problems with conditions that should only be tested based on other conditions, such as

```
i >= 0 && i < array_len && array[i] == 0 /* safe */
```

Most operators are concerned with producing values, but an assignment operator writes to its left operand as a side effect. Technically, an lvalue is an expression that refers to an object and can appear on the left side of an assignment. Examples include variables, dereferenced pointers, and elements in an array. An rvalue is an expression that produces a value and can appear on the right side. All lvalues are rvalues—their value comes from the object they refer to.

Assignment operators write to their left arguments and return the new value. This is often useful, allowing idioms like a=b=0, which sets both a and b to zero. This is also used frequently in conditionals, where a value is obtained, stored, and tested. For example,

```
while ((c = getchar()) != EOF) { /* ... */ }
```

is a common idiom for reading every character from a file; the character c is used within the body. However, this allows the common mistake of writing a = b, which assigns b to a, instead of a == b, which is true if a and b are equal.

C provides read-modify-write assignment operators. For example, l += n is equivalent to l = l + n if l does not have side effects (it is only evaluated once in l += n. Compilers often generate more efficient code for expressions involving these assignment operators, although many are now able to identify the l = l + n case.

The conditional operator b ? r1 : r2 evaluates the boolean expression b. If it is true, the operator evaluates and returns the value of r1, otherwise it evaluates and returns r2.

The comma sequence operator r1, r2 evaluates r1 first, discards its result, and evaluates r2 and returns it. This is only interesting if r1 has side effects, such as an increment or a function call. This is often used in the expressions within for loops:

```
for (i = 0, j = 100; i < 10; ++i, ++j) { /* ... */ }
```

The sizeof() operator returns the size in bytes of its argument, which much be a type. Syntactically, it behaves like a function call. It appears frequently as the argument to malloc:

```
typedef struct { int x, y; } foo_t;

foo_t *foop = malloc( sizeof(foo_t) );
foo_t *fooArray = malloc( sizeof(foo_t) * 5 );
```

Typecasting in C transforms an object of one type into another. Many transformations involve a change to the object's bit pattern, such as an int-to-double conversion, but others, such as pointer casts, usually amount to a reinterpretation.

Typecasting can be necessary, but can also cause problems when programs are compiled for different processors. One of the more popular bugs of this kind comes from casting pointers to and from integers. Pointers and integers are often the same size on 32-bit architectures but not on 64-bit architectures. Code that assumed they could be interchanged without loss of information fails on the 64-bit processor.

A few typecasts in C are always safe. For example, casting a pointer to a structure to a pointer to the type of the first element of the structure always works.

```
typedef struct { int x, y; } point_t;
point_t *p;
int *x;

x = (int *) p; /* This is safe */
```

7.5   Control Flow

Most processors provide three different ways for control to pass between instructions: directly the next instruction, directly to a distant instruction, or a conditional choice between the two. C provides restricted forms of these along with some higher-level constructs that generate more complex behavior.

In C, control can pass from instruction to instruction in six different ways. Most instructions, after they run, simply pass control to the next instruction in sequence. The goto statement sends control to a labeled statement in the same function. The if and switch statements are two- and multi-way conditionals. The three others, function call, return, and longjmp are discussed in Section 7.6.

C provides structured control-flow constructs, all of which can be built with conditionals and gotos. None are fundamental; instead, their presence helps both the programmer, who can design with more abstract structures (such as loops), and the compiler, which can more easily analyze structured control-flow and generate efficient code.

C provides few mechanisms for user-specified control constructs. Function pointers often suffice (e.g., a comparison function can be passed to a sorting function). The preprocessor (Section 7.8) can often be coerced into building new control constructs, but it is not very flexible and the results are usually fragile.

The most basic conditional is the two-way if statement. The simplest form has a single branch:

```
if (a == 0) b = 5;
```

As with all control-flow statements, a brace-delimited block can also be used, e.g.,

```
if (argc < 2) {
  sprintf(stderr, "Usage: %s pattern\n", argv[0]);
  exit(1);
}
```

Here, the statement is executed if the expression evaluates true. A variant adds a another statement executed only when the statement evaluates false:

```
if (a < 0) b = -a; else b = a;
```

which is equivalent to (note the complemented expression)

```
  if (!(a<0)) goto Else;
  b = -a; goto Next;
Else: b = a;
Next:
```

Here is a common idiom for choosing among multiple cases:

```
if (a<0) return -1;
else if (a==0) return 0;
else return 1;
```

C provides a more concise and usually more efficient multi-way conditional—the switch statement. This evaluates its expression and branches to the case label with a matching value, if there is one, to the default label otherwise.

```
switch (a) {
case 0:
case 1:  foo(); break;
case 2:  bar(); break;
default: baz(); break;
}
```

The case labels really are just labels. After the initial branch, they serve no further role. Thus, the above is equivalent to the following:

```
switch (a) {
case 0:  goto C1;
case 1:  goto C2;
case 2:  goto C3;
default: goto C4;
}
C1: ;
C2: foo(); goto Break;
C3: bar(); goto Break;
C4: baz(); goto Break;
Break:
```

C provides three looping constructs. The do-while statement repeats a statement while a predicate expression is true. The body always runs at least once.

```
do {
  foo();
  bar();
  --a;
} while (a > 0);
```

is equivalent to

```
Again:
  foo(); bar(); --a;
if (a>0) goto Again;
```

The while construct runs the test first.

```
while (i>0) {
  foo();
  --i;
}
```

is equivalent to

```
  goto Test;
Again:
  foo(); --i;
Test:
  if (i>0) goto Again;
```

The for construct adds expressions that run before the loop begins and after the statement completes, generally used for loop initialization and increment respectively.

```
for (i=0 ; i<10 ; i++) {
  foo(i);
}
```

is equivalent to

```
  i=0; goto Test;
Again:
  foo(i); i++;
Test:
  if (i<10) goto Again;
```

Infinite loops are written

```
for(;;) { /* ... */ }
```

An infinite loop usually has a way to escape. A return or goto suffice, but C also provides the structured break to leave the inner loop and continue to restart it.

```
for (;;) {
/* Continue: */
 if (i == 10) break;  /* goto Break */
 if (a < b) continue; /* goto Continue */
 }
 /* Break: */
```

7.6   Functions

A C program is ultimately a collection of functions—C's main high-level construct. These provide a mechanism for grouping, data hiding, storage management, a divide-and-conquer style of design, and recursion. Each function has a name, an interface (the number and type of its arguments and a return type), and a body consisting of statements. Calling a function passes values for each of the arguments its the body, which processes them and returns a value.

C's functions resemble their mathematical counterparts. The syntax for calling a two-argument function f is what you would expect: f(x,y). However, a C function may have side-effects: it may modify global variables or a data structure passed to it as a pointer. So each call of f(1,2) may return a different value.

C's function-centric view encourages a divide-and-conquer style of design where large tasks are divided into many smaller tasks. The view is imperative: it encourages thinking of a program as series of things to do, such as copy this to that, put this information in that data structure, and so forth.

C uses a stack to handle arguments, automatic variables, and return addresses for function calls. As an example, consider calling

```
int min(int a, int b)
{
   return a < b ? a : b;
}
```

in the expression 2 + min(x+3,y). First, the arguments x+3 and y are evaluated and pushed on the stack. Next, a return address that will bring control back to code for evaluating this expression is pushed on the stack and the function is called. The function evaluates a<b?a:b using the values of a and b on the stack, and stores the result in a processor register (the effect of the return). The return address is popped from the stack and jumped to. Control resumes in code for finishing the evaluation of the expression involving min, which adds 2 to the result register to produce the result.

Using a stack allows any function to call any other without fear of local variable storage colliding because each call uses new space on the stack. This also permits a function to call itself either directly or indirectly. Such recursion is a powerful technique for solving problems on self-referential data structures such as graphs.

Accessing arguments and local variables on the stack requires accessing memory relative to the top of the stack. This does imply one more addition than using a global address, but this can actually be faster since stack offsets are often small integers; loading the wider global address can be slower and larger.

C uses call-by-value semantics: a called function gets its own local copy of its arguments and it is not able to change them. This has the advantage of making function calls safer in the sense that they are not usually able to modify data in their calling function. For example, the obvious way to write a "swap" function does not work:

```
void swap(int x, int y) /* This does not work */
{
  int tmp = x;
  x = y;
  y = tmp;
}
```

(C++ references would make this work. See Section 8.1.)

Pointers can solve this problem. Instead of passing values, passing pointers to them allow them to modified. For example, calling the following with swap(&a, &b) passes the addresses of a and b and swaps their values.

```
void swap(int *x, int *y)
{
  int tmp = *x;
  *x = *y;
  *y = tmp;
}
```

The return statement is the natural way to exit a function and return a value, but there are times, such as when an unexpected error occurs deep within nested functions, that it is convenient to leave many functions at once. The standard library's setjmp and longjmp can do this: exit from arbitrarily nested functions.

Setjmp saves a point in the program that longjmp can later return to. Setjmp returns zero when it is first called. Later calling longjmp with the buffer setjmp prepared returns control to the setjmp. This also returns the integer passed to longjmp, allowing an error code to be passed up. A typical use:

```c
#include <setjmp.h>

jmp_buf jmpbuf; /* Storage for the return location */

void top(void)
{
  switch (setjmp(jmpbuf)) {
  case 0:
    subfunction();        /* Called the first time */
    break;
  case 1:
    error_condition(); /* longjmp() returns here */
    return;
  }
}

void subfunction(void) { deeplyNested(); }

void deeplyNested(void)
{
  /* ... */
  longjmp(jmpbuf, 1); /* Return with error code 1 */
}
```

This is a fragile mechanism that modifies the stack without the compiler's knowledge, and can cause problems when optimization is turned on. For example, if setjmp is part of a large expression, there is no guarantee the expression will be evaluated correctly when longjmp returns, since evaluation may be partially complete where setjmp occurs. Furthermore, any automatic variables declared in the setjmp function may not have their values restored after the longjmp because they were placed in registers and not restored.

7.7 The Standard Library

The C standard library, which most implementations support, contains a collection of functions, macros, and constants that provide file-oriented I/O, common mathematical functions (sin, cos), string handling, and memory management (malloc, free).

Header	Description	Example
<assert.h>	Generate runtime errors	assert(a > 0)
<ctype.h>	Character classes	isalpha(c)
<errno.h>	System error numbers	errno
<float.h>	Floating-point constants	FLT_MAX
<limits.h>	Integer constants	INT_MAX
<locale.h>	Internationalization	setlocale(...)
<math.h>	Math functions	sin(x)
<setjmp.h>	Non-local goto	setjmp(jb)
<signal.h>	Signal handling	signal(SIGINT,&f)
<stdarg.h>	Variable-length arguments	va_start(ap, st)
<stddef.h>	Some standard types	size_t
<stdio.h>	File I/O, printing.	printf("%d", i)
<stdlib.h>	Miscellaneous functions	malloc(1024)
<string.h>	String manipulation	strcmp(s1, s2)
<time.h>	Time, date calculations	localtime(tm)

Figure 7.3: The header files of the C standard library

C programs almost always use the I/O library, which can operate on files or steams. Common operations are opening and closing files (fopen, fclose), reading and writing characters (fgetc, fputc), and reading and writing formatted strings (fscanf, fprintf).

7.8   The Preprocessor

C's preprocessor performs macro substitution, conditional compilation, and inclusion of files. Virtually all programs use the preprocessor to include header files containing extern declarations of functions and variables used throughout a large program. For example,

```
#include <stdio.h> /* Include I/O library declarations */
#include "foo.h"   /* Include my own declarations */
```

Defining symbolic constants is another common use. This is a simple text substitution mechanism.

```
#define BUFSIZE 1024
char buffer[BUFSIZE];
```

Macros may take textual arguments, making their use look like a function call, but it is important to remember they only represent simple text substitutions.

Function-like macros can cause problems. Precedence rules can cause expressions passed as arguments to be interpreted in surprising ways; parenthesizing each argument in the macro's definition along with the definition itself solves this problem. Arguments with side effects can also produce unexpected results. Unlike a function call, which always evaluates its arguments exactly once, a macro argument may be evaluated, zero, one, or more times depending on the definition. For example, if

```
#define min(a,b) ((a) < (b) ? (a) : (b))
```

is used as min(++x,++y), x and y will always be incremented once, but the smaller of the two will be incremented twice. The assert macro, part of the standard library, exhibits this problem. If debugging is turned off, it expands to nothing, otherwise it evaluates its expression and halts the program if it evaluates false. An assert expression with side-effects can lead to a hard-to-diagnose bug, since it only appears with debugging disabled.

Conditional compilation is the other common use of the preprocessor. The #ifdef conditional controls whether the code between it and a matching #else or #endif makes it to the compiler. This is often used to hide debugging code:

```
    a = b + c;
#ifdef DEBUG
    printf("a: %d\n", a);
#endif
```

Machine dependencies is a common application:

```
#ifdef SPARC
#   include <sparc/a.out.h>
#else
#   include <a.out.h>
#endif
```

7.9   History

As Dennis Ritchie explains [69], C was developed mostly during the period of 1971 to 1973 at Bell Telephone Laboratories in parallel with the development of the Unix operating system. It evolved from the B language, a stripped-down version of BCPL. All three were designed for systems programming (the development of operating systems and associated tools), were "close" to the hardware, and were designed to operate on small machines.

B and BCPL were both typeless languages that only manipulated machine words. This was a close match to the hardware since these languages ran on word-addressed machines. Treating a number as a memory address was transparent and an operation that worked on a number would work equally well on an address. To this day, C retains this powerful blurring between integers, characters, and addresses.

The B compiler generated threaded code: a sequence of addresses of primitive routines. An interpreter calls these in sequence to run the program, making for slow but compact code.

String-handling was problematic in such an environment. Library routines, coded in assembly, could be used for certain manipulation, but more complex operations involved unpacking a string into one word per character, operating on it, and repacking the results.

C came about from the need to handle variable-sized data types, improve code speed, and deal with the new byte-addressed PDP-11. Characters joined integers as native types, along with pointers and arrays of both. Address arithmetic continued to work, but the expression p + 3 would do different things depending on whether p was a character or integer pointer.

Ritchie considers the relationship between arrays and pointers the crucial jump between BCPL and C. In BCPL, int a[10]; meant space for a pointer called a was created and initialized to point to the base of an array of ten integers. These semantics created problems when arrays appeared in structures. Consider a structure that might describe an early Unix directory entry:

```
struct dirent {
  int  inumber;
  char name[14];
};
```

Allocating an array of these would require initializing each name field pointer. This would be possible but costly for explicit arrays, but handling arrays allocated using malloc would be very complicated. (C++ objects can be given these semantics, but only because the language supplies the new operator that knows when arrays are allocated.)

The solution was to create the pointer when it was used instead of storing it in memory. The declaration int a[10]; now allocates space for ten integers. In an expression, a produces the address of this space: a constant for a global variable, an offset from the top of stack for an automatic, or an offset from the base address of a struct.

Concretely, if map is defined as

```
struct m {
  char name[10];
  char alias[20];
} map[5];
```

then map[3].alias returns the sum of the base address of map, three times sizeof(struct m), and the offset of alias.

When ANSI standardized the language in the late 1980s, the biggest change was to add argument types to function prototypes. In the earlier version of the language (usually called K&R C after the first edition of Kernighan and Ritchie's book [48]), function declarations (i.e., forward or extern declarations) only defined a function's return type. It assumed all arguments were the same size, a holdover from BCPL's exclusive use of machine words. ANSI allows a program to supply the type of each argument, permitting arguments of variable widths and making it possible to check types.

Old-style declarations remain for compatibility but cause a problem: how can a new-style definition with zero arguments be distinguished from an old-style definition? The solution is ugly: a new-style definition with zero arguments is written with a single void argument. Concretely:

```
int foo(i, c)         /* old-style definition */
int i;
char c;
{ /* ... */ }

int bar(void) { /* ... */ } /* new-style */
int baz(int i, char c) { /* ... */ } /* new-style */
```

The C preprocessor was added relatively late, explaining its less-than-perfect integration. The first version had little more than the ability to #include files and #define parameterless macros.

The earliest C parsers were hand-written recursive descent. This has left a few blemishes on C's syntax that make it difficult to write a bottom-up LALR(1) parser (e.g., YACC-style [1]).

7.10   Compilation

The C language was designed to allow large programs to be compiled in pieces. A compiler needs little global information to compile a function; everything that is needed must have been defined earlier, allowing allow single-pass compilation. A file only needs external declarations for the global variables and functions it uses. A compiler can even compile a function in pieces since the only global information, the automatic variables, are declared before they are used. Modern optimizing compilers prefer instead to analyze large portions of a program to produce better code.

Generating code for an expression is the basic challenge in compiling C. Doing it correctly is fairly easy: translate the expression into a tree and walk it depth-first, generating a small piece of code for each node that operates on a stack. A constant becomes code that pushes its value on the stack; a reference to a variable pushes its value on the stack; and an operator like + pops the top two numbers on the stack, adds them, and pushes the result.

Many techniques can make expression code more efficient. For example, constant propagation simplifies constant expressions, e.g., replacing 2+2 with 4. Using registers instead of the stack is often more efficient. Another optimization replaces groups of basic operations (such as adding a constant to a result) with fewer or one processor instruction. Further optimization often comes from considering how data flows between sequences of expressions. In general, better understanding of the program allows a compiler to produce better code.

Generating code for conditional and loop instructions is easy. The goto statement (and any implicit in other statements) becomes an unconditional branch. The if statement becomes an expression evaluation followed by a conditional branch.

Compilers often generate interesting code for switch statements. When the cases are few or sparse, the generated code is usually just a sequence of simple tests, but when the cases are densely packed, many compilers generate code that jumps to an address in a table indexed by the result of the expression. The most efficient way to do this depends on the distribution of the case labels, the instruction set of the processor, and the compiler's desire to reduce time or space.

Code for a function begins by allocating space on the stack and ends by releasing this space and returning to the caller.

Large programs are compiled by a code generation step followed by linking. Code generation produces assembly code that may contain unresolved references to external functions and variables. The linker collects the code generated from each file, examines the global functions and variables provided by each, and uses this information to resolve these references.

C links global functions and variables by name, so the linker cannot catch inconsistent declarations with the same name, allowing

```
int foo(int, char);          /* in file A */
extern char foo(struct box);  /* in file B */
```

to slip through. The program will compile and link successfully but fail when the function is called since the caller does not send the type of arguments the callee expects. A program called lint checks such cross-file consistency and would report this error.

Many books have been written on compilers for C-like languages. Aho, Sethi, and Ullman's "Dragon Book" (from its cover illustration) [1] is the classic reference. Fraser, Hanson, and Hansen's book on their lcc compiler [30] is easily the most in-depth description of a C compiler ever written. Muchnick's book [61] describes many of the advanced techniques used in modern optimizing compilers.

7.11 Alignment of Structures, Unions

Because many processors require wide data to be aligned (i.e., data wider than a byte must be stored at addresses that are multiples of its width), data within structures and unions is usually padded. Both structures and unions must array correctly, that is, alignment restrictions must be obeyed when two copies of the same structure or union are placed in memory one after the other. So their size is not necessarily the sum of the sizes of their members, and rearranging the order of fields in a structure can affect its size.

For example,

```
sizeof(int)                                         =  4
sizeof(char)                                        =  1
sizeof(struct { char a[5]; })                       =  5
sizeof(struct { int a; char b; })                   =  8
sizeof(struct { int a; char b; int c; char d; }) = 16
sizeof(struct { int a; char b; char d; int c; }) = 12
```

Unions are large enough to fit the largest thing they may contain, but alignment restrictions may force them to be larger.

```
sizeof(union { int a; } )            = 4
sizeof(union { char a[3]; } )        = 3
sizeof(union { char a[5]; } )        = 5
sizeof(union { int a; char b[3]; } ) = 4
sizeof(union { int a; char b[5]; } ) = 8
```

7.12 Nondeterminsm

A compiler is free to implement many things in C as it wishes for efficiency. A compiler usually chooses the sizes of the built-in types based on what generates the fastest code, often causing problems when code that assumes these sizes is ported to different platforms.

Function arguments may be evaluated in any order, so

```
int a = 0;
printf("%d %d %d\n", ++a, ++a, ++a);
```

might print 1 2 3 or 3 2 1. This is generally not a problem: a programmer should avoid things with interacting side-effects in function arguments. This example could easily have been made deterministic:

```
int a = 0;
printf("%d %d %d\n", a+1, a+2, a+3);
a += 3;
```

Like most forms of nondeterminism, this makes it harder to write correct, portable programs. Testing it on one platform is no guarantee of correctness.

7.13   Exercises

7–1.   (a) Why did C's designers make the "&&" and "||" operators "short-circuit"? (b) What assembly language do they produce?

7–2.   Why and when is a multi-way switch-case statement be more efficient than a series of if-thens?

7–3.   How would you mimic the effect of setjmp-longjmp using other C statements if this functionality wasn't provided by the standard library? Hint: it would require significant changes to a program.

7–4.   Why did ANSI C adopt the newer style of declaring function arguments? Why was it not necessary earlier? What errors did the change help programmer to avoid?

7–5.   Write a small C program that tests the order in which function arguments are evaluated by your compiler. Does the compiler's -0 flag change this order?

8

C++

C++ is a language built on C that supports object-oriented programming as well as a macro style of polymorphism. Bjarne Stroustrup, who created the language, wanted to provide more mechanisms for structuring large programs while maintaining the efficiency of generated code. The result is a much larger, more structured language with a stronger, more elaborate type system.

C++'s constructs encourage encapsulation and code reuse. A new type (class), such as a rectangle, can be defined by specifying its representation and the operations (methods) that can be performed on it. A new class (say, a filled rectangle) can be derived from an existing one by adding methods. Templates provide macro-like polymorphism, e.g., a sort algorithm can be written to work with integers or floating-point numbers. The compiler generates custom object code for each use of the sort algorithm.

C++ is a superset of C. A C++ compiler can compile C programs, and a C++ program can call C object code. This evolutionary path sped the adoption of the language since existing C libraries and code continued to be useful.

The C++ standard library includes a collection of polymorphic data structures often referred to as the standard template library or STL. The STL provides arrays, binary trees, and linked lists that are often as efficient as custom-crafted implementations thanks to C++'s template facility, which customizes the code for each data type a template is applied to. The C++ library also includes a powerful string class that largely frees a programmer from having to deal with string memory management.

Support for C++ is less stable than that for C. Part of this is due to it being a younger language, but it is also a very large language that was only recently (ca. 1998) standardized. Support for newer language features like namespaces and parts of the template facility remains spotty in many compilers. The standard template library taxes most compilers; some only provide partial support.

Three books by Stroustrup are among the best references on the language. The best starting point is the third edition of *The C++ Programming Language* [73]. This is a large book that covers all of C++ in a style that is both tutorial and reference. The *Annotated C++ Reference Manual* [28] goes into greater detail but does not discuss the standard library. Finally, *The Design and Evolution of C++* [72] recounts both the history and motivation for why C++ is like it is. Unlike the other books, it also discusses implementation issues.

The object-oriented style of programming is powerful and easier to misuse than C's function-oriented style. Gamma et al.'s book *Design Patterns* [31] describes some excellent ideas for structuring large object-oriented programs.

8.1   Overview

C++ adopts virtually all of C's syntax and extends it.

New variables may be declared anywhere within the body of a function, not just at the beginning as in C. Avoiding uninitialized variables is the main advantage here, since the declaration of a variable can usually wait until its initial value is ready.

Declaring a loop's index variable within the for statement is particularly convenient. C++ allows loops to be written

```
for (int i = 0 ; i < 10 ; ++i ) { /* ... */ }
```

Unfortunately, the scope of the variable i has had two interpretations. In earlier versions of C++, the scope of i extended beyond the body of the loop, prohibiting code like

```
for (int i = 0 ; i < 10 ; ++i ) { /* ... */ }
for (int i = 10 ; i > 0 ; --i ) { /* ... */ }
```

since the two definitions of i would collide. The latest version of the language restricts the scope to the body of the loop, but for the moment, many compilers either misinterpret or give a warning about the new usage.

C++ expands C's ability to call functions by adding default arguments and overloading based on argument types. For example,

```
void foo(int) {}                   // foo1
void foo(int, int, int = 0) {}  // foo3

void main()
{
   foo(1);      // call foo1(1)
   foo(1,2);    // call foo3(1,2,0)
   foo(1,2,3); // call foo3(1,2,3)
}
```

C++ provides a reference type as an alternative to a pointer. This makes it possible, for example, to implement an elegant function that swaps two integers:

```
void swapInt(int& a, int& b)
{
  int tmp = a;
  a = b;
  b = tmp;
}

main()
{
   int x = 5;
   int y = 6;
   swapInt(x, y);
   // x = 6 and y = 5 here
}
```

A reference sets up an alias to the given argument (i.e., a pointer). When swapInt is called, the compiler passes the addresses of x and y to the function, which treats them accordingly.

A const reference behaves like a call-by-value argument but avoids a copy. This can be much more efficient with a large object.

8.2   Classes

Programs are often conceived of as a collection of data structures (classes) on which certain operations (methods) can be performed. C++ directly supports this style.

Dividing a program into classes and methods can make it more robust. Maintaining consistent data is easier since what can be done to it is limited. The barrier between the representation of an object and its interface also simplifies program maintenance: the representation can easily be changed without affecting the rest of the program.

In C++, a class definition defines a new type of object. It defines how objects of the new type are represented and the operations that may be performed on them.

A class definition is an extended C structure definition. In addition to defining named fields of objects of the class, it may also include member functions, constructors and destructors, and static fields and member functions. A class definition may contain type definitions.

Consider a rectangle type:

```
class Rectangle {
    int x, y, h, w;    // Location and size
    static int count; // Total number of rectangles
public:
    Rectangle(int, int, int, int);    // Constructor
    ~Rectangle();                     // Destructor
    int area() const { return h * w; }
    void move(int, int);
    void draw() const;
    static int total() const;
};
```

The members x, y, h, and w are private data because they fall before the public: directive. Each Rectangle object will have its own copy of them, and only member functions may access them.

The member count is marked static to associate it with the class rather than objects of the class. Think of it as a global variable associated with the class. It, too, is only visible to member functions, but unlike x, which can be initialized by the constructor, count must be defined and initialized somewhere, e.g.,

```
int Rectangle::count = 0;
```

The `Rectangle` member is the class's constructor, a function that initializes a new object of the class. For example:

```
Rectangle::Rectangle(int xx, int yy, int hh, int ww)
{
  x = xx; y = yy; h = hh; ww = w;
  ++count;
}
```

With this definition, a new rectangle can be created in different places, but any constructor arguments must be supplied.

```
Rectangle re(10,0,30,35); // Create a variable
// Rectangle rbad;         an error: missing arguments

// Create a new object in dynamic store
Rectangle* myrect = new Rectangle(5,8,20,21);

// Call the destructor and free memory
delete myrect;
```

C++ provides a shorthand for initializing members of a class in a constructor. For example, here is an equivalent `Rectangle` constructor:

```
Rectangle::Rectangle(int xx, int yy, int hh, int ww)
  : x(xx), y(yy), h(hh), w(ww)
{
  ++count;
}
```

The `˜Rectangle` member is the destructor, a member function called whenever an object of the class falls out of scope or is deleted. A destructor is responsible for cleaning up after the object, often by releasing resources the object acquired during its lifetime. The destructor is not responsible for releasing memory used for the object itself—the compiler provides code that does this automatically. For the rectangle class, the destructor updates the global rectangle count:

```
Rectangle::˜Rectangle()
{
  --count;
}
```

The `area` function is marked `const` to indicate it does not modify the object. The compiler will complain if a `const` member attempts to change the object, or if a non-`const` member function is called on a `const` object. This is a powerful mechanism for avoiding bugs.

Supplying the body of the `area` member within the class definition indicates it is to be inlined. That is, instead of calling the function, the compiler will copy the function's body when it is called. Such function inlining greatly speeds small functions, but for large functions the speed advantage is negligible and can greatly increase code size.

The `area` member also illustrates how member functions access their object's data members. The reference to h in `area`'s body refers to the h member of the object on which `area` was called. In effect, member functions are called with an implicit parameter named `this` that is a pointer to the object. Thus this definition of `area` behaves as if it were written in C as

```
int Rectangle_area(const Rectangle *this)
{
    return this->h * this->w;
}
```

The `draw` method might draw the rectangle on some output device. It is marked `const` since it should not change the rectangle.

The `move` method is straightforward:

```
void Rectangle::move(int dx, int dy)
{
    x += dx; y += dy;
}
```

The `total` member function is `static`, meaning it is is called directly and does not operate on an object of the class. As a result, it may only access other `static` data members of the class. Its definition:

```
int Rectangle::total()
{
    return count;
}
```

Member functions are invoked using a structure-like syntax:

```
Rectangle re(10,0,30,35);
int area = re.area();

Rectangle* rep = new Rectangle(5,3,1,2);
rep->move(2,3);
```

Static member functions are called using the : : notation to specify the class name. This notation is not necessary for non-static members since the object type includes the class name.

```
int current_number = Rectangle::total();
```

8.3   Single Inheritance and Virtual Functions

Programs often use many similar types. A program using Rectangle might also want Square and Circle.

Similar classes can be built using inheritance. A derived class is built by adding functionality and data to a base class.

Here is an example. The Shape class defines a basic shape with an origin that may be moved. Rectangle and Circle classes are derived from it.

```
class Shape {
   int x, y;    // Origin
public:
   Shape(int xx, int yy) : x(xx), y(yy) {}
   void move(int dx, int dy) { x += dx; y += dy; }
   virtual int area() const = 0;
};
```

The area method is marked virtual, meaning it may behave differently for different types of shapes. It is also marked pure virtual (= 0), indicating the Shape class itself does not have a definition for area. This also makes the Shape class abstract, meaning no Shape objects may be created.

Virtual functions make it much easier to add a new type of object to a collection. In C, the same behavior is often written with a big switch-case statement. For example, a graphics program might have a function that draws each shape in a list that first checks the shape type before drawing it. C++'s virtual functions replace such structures, placing responsibility for different operations with the objects instead of the rest of the program. Adding a new shape only requires a new type to be defined. The default draw method for the new type will be used unless it is explicitly overridden. A draw-all-shapes function does not need to be changed.

The Rectangle and Circle classes are also Shapes. They inherit all the members of Shape except for constructors and destructors. So both Rectangle and Circle objects have x and y members, and the move method may be called on either one.

```
class Rectangle : public Shape {
  int h, w;
public:
  Rectangle(int xx, int yy, int hh, int ww)
    : Shape(xx, yy), h(hh), w(ww) {}
  int area() const;
};

class Circle : public Shape {
  int r;
public:
  Circle(int xx, int yy, int rr)
    : Shape(xx, yy), r(rr) {}
  int area() const;
};

int Rectangle::area() const { return h * w; }
int Circle::area() const { return 3 * r * r; }
```

Both Rectangle and Circle define their own area member functions. These may be called like this:

```
Shape* s = new Rectangle(10,10,20,30);
int area1 = s->area(); // Invoke Rectangle::area()
delete s;

s = new Circle(10, 10, 15);
int area2 = s->area(); // Invoke Circle::area()
delete s;
```

Using virtual functions adds some overhead: a word to each object and an extra lookup for each function call (Section 8.11). This is usually a reasonable cost to pay for the power of virtual functions.

8.4   Multiple Inheritance

C++ allows multiple inheritance, where a class is derived from two or more bases. For example, colored shapes might be added by defining classes as follows:

```
class Colored { /* ... */ };
class CRectangle : public Rectangle, public Colored
{ /* ... */ };
class CCircle : public Circle, public Colored
{ /* ... */ };
```

Multiple inheritance can easily lead to ambiguity. Members that appear in more than one base class must be resolved.

```
class B1 { protected: int a; };
class B2 { protected: int a; };
class C : public B1, public B2 {
public:
  int geta() const {
    return B1::a + B2::a;
  }
};
```

More confusion can arise when base classes are derived from a common one.

```
class A { /* ... */ };
class B1 : public A { /* ... */ };
class B2 : public A { /* ... */ };

// class C2 has two copies of A
class C2 : public B1, public B2 { /* ... */ };
```

Here, C2 contains two copies of A, disambiguated with B1::A:: and B2::A::. C++ also makes it possible for B1 and B2 to share the A they inherit by declaring their inheritance of A virtual:

```
class B3 : virtual public A { /* ... */ };
class B4 : virtual public A { /* ... */ };

// class C1 has a single copy of A
class C1 : public B3, public B4 { /* ... */ };
```

Constructing an object containing virtual base classes can be complicated. Generally, an object is responsible for passing arguments to the constructors of the classes from which it is derived. Because the language does not allow objects to be constructed twice, this leads to a dilemma for virtual base classes. Who is responsible for constructing them? Classes derived from them cannot, since two or more would try to invoke the constructor.

C++ solves this problem by either allowing a single derived class to call the virtual base class's constructor explicitly, or implicitly calling it with no arguments.

For example,

```
struct A {
  int a;
  A(int aa) : a(aa) {}
  A() : a(0) {}
};

struct B1 : virtual public A { B1(int a) : A(a) {} };
struct B2 : virtual public A { B2(int a) : A(a) {} };

struct C : public B1, public B2 {
  // In both cases, B1::B1(int) and B2::B2(int)
  // do not call A::A(int)

  // Calls A::A(), B1::B1(int), and B2::B2(int)
  C(int b1, int b2) : B1(b1), B2(b2) {}

  // Calls A::A(int), B1::B1(int), and B2::B2(int)
  C(int a) : A(a), B1(0), B2(0) {}
};
```

8.5 Namespaces

Name clashes are a common problem when large programs are assembled from existing pieces. For example, two pieces might both define a function called print. C++ classes already help this problem by allowing two methods on different objects to be named print with no danger of ambiguity. But class names are global by default.

A C++ namespace groups classes, types, variables, and functions into a separate scope, removing them from the global environment.

The effect is like a class definition, but it is not possible to create an object whose type is the namespace. A namespace does have the notion of a member function.

You might place the Shape classes in their own namespace:

```
namespace Shape {
  class Shape /* ... */;
  class Rectangle : public Shape /* ... */;
  class Circle : public Shape /* ... */;
}
```

Outside the namespace, its members can be referred to using the :: notation:

```
Shape::Shape* s = new Shape::Rectangle(10,10,20,30);
```

Alternately, single declarations may be imported into the current scope with the using directive:

```
using Shape::Shape;
using Shape::Rectangle;
Shape* s = new Rectangle(10,10,20,30);
```

The contents of an entire namespace may also be imported:

```
using namespace Shape;
Shape* s = new Circle(10, 10, 15);
```

Like class members, namespace members must be declared within the namespace, but they may be defined inside or outside. Unlike classes, defining a functions within a namespace does not mean to inline the function.

```
namespace Both {
  int foo() { /* ... */ }
  int bar();
}

int Both::bar()
{
  // ...
}
```

Namespaces are open: members may be added after the first definition:

```
namespace Mine {
  int foo() { /* ... */ }
}

namespace Mine {
  int bar() { /* ... */ }
}
// Mine now has both foo() and bar()
```

Namespaces with long names are generally a good idea, since they avoid the nasty problem of namespace name clashes. However, long names become cumbersome in code. C++ provides a way to alias the name of a namespace to avoid this problem:

```
namespace Very_Long_Namespace_Name {
  int foo() { /* .. */ }
}

Very_Long_Namespace_Name::foo();

namespace VLNN = Very_Long_Namespace_Name;

VLNN::foo();
```

Namespace aliases may also be used to easily switch between different implementations of the namespace. For example,

```
namespace Lib_old { /* ... */ }
namespace Lib_new { /* ... */ }

namespace Lib = Lib_new;
Lib::member();
```

8.6   Templates

Types in C and C++ are a double-edged sword. On the one hand, they can catch errors by preventing inappropriate operations, such as integer arithmetic on a floating-point number. On the other, they often prevent legitimate code reuse.

C++'s virtual functions enable some algorithm reuse. For example, a general sorting algorithm might be written like this:

```
class Sortable {
  virtual bool lessThanEqual( const Sortable& ) = 0;
};

void sort( int, Sortable* )
{
  /* Sort an array of Sortable objects */
}

struct SortableInt : public Sortable {
  int val;
  bool lessThanEqual( const Sortable& s ) {
    return val <= s.val;
  }
};
```

This approach has two disadvantages. First, all objects to be sorted must be derived from the Sortable base class. An object on which multiple algorithms may be applied would have to inherit from multiple abstract interface classes, requiring multiple inheritance and its associated overhead and confusion. Second, this mechanism also requires a virtual function call for each operation. The overhead for this, especially for something as simple as a comparison, can easily exceed the cost of the operation, making such code very inefficient.

C++'s templates provide an alternative mechanism for reusing algorithms and classes based on the idea of macro expansion. The C preprocessor can be used a primitive mechanism to do this. Consider a function that returns the minimum of two numbers. For integers, this would be written

```
int fmin(int x, int y) { return x < y ? x : y; }
```

The only difference for floating-point number would be to change the ints to doubles, suggesting it could be defined in C with a macro:

```
#define mmin(x,y) ((x) < (y) ? (x) : (y))
```

The parenthesis around the arguments avoid unexpected results caused by operator precedence. Unfortunately, mmin(++a,b) increments a twice if a is the smaller of the two arguments.

In C++, the min operation is best written with a template:

```
template <class T> T& min(const T& a, const T& b)
{
  return a < b ? a : b;
}
```

When min is called, the compiler recognizes the types of its arguments and instantiates the appropriate template. For example,

```
int mi, xi, yi;
mi = min(xi, yi); // Use min(int, int)

double md, xd, yd;
md = min(xd, yd); // Use min(double, double)
```

A much better generalized sorting routine would be written using templates like this:

```
template <class T> void sort( int size, T* ar)
{
  // ...
  if ( ar[i] > ar[j] ) {
    T tmp;
    tmp = ar[i];
    ar[i] = ar[j];
    ar[j] = tmp;
  }
  // ...
}
```

When a compiler encounters a call to sort, it may create a new version of the sort function specialized to a new type. The effect resembles macro expansion, but the function is not inlined. A macro with a type as an argument is a crude way to do this in C:

```
#define defSort(T) \
void sort_##T( int size, T* ar) \
{ \
  T tmp; \
  /* ... */ \
}

defSort(int);   /* Defines sort_int() */
defSort(float); /* Defines sort_float() */
```

Although a template does not explicitly list the demands placed on its type arguments, the compiler will generate an error message if the type does not provide the operations used in the template. Such errors can be fairly cryptic, since they occur within template code, but this eliminates the need to derive arguments from special classes. An algorithm like sort is just as applicable to arrays of integers as arrays of a user-defined type.

The standard library provides an even more flexible sorting algorithm that can sort anything that behaves like an array of things that can be compared.

8.7 Exceptions

C++ provides exceptions, a high-level mechanism intended primarily for error handling. Intended as a replacement for C's very low-level setjmp/longjmp (see Page 128), it allows a deeply-nested function to throw an exception that returns control to an exception handler in one of its calling functions.

Exceptions are used like this:

```
struct My_Exception {
  int i;
  My_Exception(int ii) : i(ii) {}
};

void bar() {
  // ...
  if ( problem ) throw My_Exception(3);
}

void foo() {
  try {
    bar();
    // ...
  }
  catch (My_Exception e) {
    cerr << "Caught My_Exception(" << e.i << ")\n";
  }
}
```

Exceptions are often put in a class hierarchy so related ones can be handled cleanly. For example:

```
struct Math_Ex { /* .. */ };
struct Divide_Ex : public Math_Ex { /* .. */ };
struct Overflow_Ex : public Math_Ex { /* .. */ };

void foo() {
  try {
    // ...
  }
  catch (Divide_Ex e) {
    // Handle division specially
  }
  catch (Math_Ex e) {
    // Handle all other math exceptions
  }
}
```

C++'s exception facility guarantees all local variables defined between throw and catch are destroyed just as they would be using the normal function return mechanism.

However, this only applies to local objects. Those allocated with new may be lost if care is not taken.

To assist with the deallocation of such objects, the C++ standard library provides the auto_ptr template. This behaves like a normal pointer except that the object it points to is automatically deleted when it falls out of scope, e.g., when an exception is thrown.

C++ provides another facility for cleaning up when an exception is thrown. Writing catch(...) catches all exceptions. Once this is done, clean-up code can run, and the exception can be re-thrown by simply calling throw without any arguments.

8.8   Operator Overloading

C++ allows the behavior of the standard arithmetic operators (i.e., those listed in Figure 7.2 on Page 121) to be defined for user-defined types. A complex number class might override the + operator like this:

```
struct Complex {
  double r, i;
  Complex(double rr, double ii) : r(rr), i(ii) {};
};
```

```
Complex operator+(const Complex& a, const Complex& b)
{
    return Complex( a.r + b.r, a.i + b.i );
}

main()
{
    Complex a(10,20), b(30,40);
    Complex c = a + b;
}
```

8.9   The Standard Library

The standard library provides stream-based I/O, a string class, mathematical functions and a mathematical vector class, polymorphic container classes with iterators, and standard algorithms such as sorting that can be applied to the containers.

C++ I/O is built on the concept of a stream: an unlimited sequence of characters. Streams can be files, input or output to a terminal, network connections, or even strings in memory.

Output uses the << operator overloaded to take a stream on the left and return a stream. All built-in types overload this operator to provide type-specific output, and it is easy to extend this to user-supplied types. One of the standard output streams is called cout. It can be used like this:

```
cout << "Hello World\n";
```

A complex number class might provide this:

```
ostream& operator<<(ostream& s, const Complex& c) {
    return s << c.r << " + " << c.i << 'i';
}

complex c(5,7);
// Prints "The value of c is 5 + 7i"
cout << "The value of c is " << c << '\n';
```

Input is done similarly. It uses the >> operator with an input stream on the left and a reference to the object to read on the right. The standard input stream is named cin. An example:

156   C++

	Containers
`<vector>`	Variable-size 1-D array
`<list>`	Doubly-linked list
`<queue>`	Queue
`<deque>`	Double-ended queue
`<stack>`	Stack
`<map>`	Ordered associative array
`<set>`	Set
`<bitset>`	Set of booleans
	String-related
`<string>`	Strings
`<cwtype>`	Wide character classification
`<cwchar>`	wide string manipulation
	I/O
`<iostream>`	Basic I/O stream facilities
`<streambuf>`	Stream buffers
`<ostream>`	Output stream
`<istream>`	Input stream
`<iomanip>`	Stream manipulators
`<sstream>`	String streams
`<iosfwd>`	Forward declarations
`<ios>`	iostream bases
	Numerical
`<limits>`	Numeric limits
`<complex>`	Complex numbers
`<valarray>`	Numeric vectors
`<numerics>`	Numeric operations
	General
`<utility>`	Operators and pairs
`<functional>`	Function objects
`<new>`	Memory management
`<memory>`	Allocators for containers
`<iterator>`	Iterators
`<algorithm>`	Algorithms: search, sort
`<stdexcept>`	Standard exceptions
`<typeinfo>`	Runtime type identification
`<exception>`	Exception handling support
`<locale>`	Cultural localization

Figure 8.1: The header files of the C++ standard library, which also includes the C standard library (Figure 7.3).

```
int i;
cin >> i;   // Read an integer and save it in i
```

The standard string class is particularly convenient because it relieves the programmer of managing memory for strings. They operate as you might expect:

```
string s; // s is now the empty string;
s = "Hello";
cout << s << '\n';
s += " everybody";    // s becomes "Hello everybody"
s = "Replaced";       // releases memory for the old s
string j = "I was " + s;
```

The string class also provides comparison operators, the ability to extract and insert substrings, and find and replace operations.

The standard library's container classes are built using templates, allowing them to contain virtually any class without loss of efficiency. They include the array-like vector that provides constant-time indexing; the doubly-linked list; stack, queue, and deque; a set; and the associative array map. The set and map containers require the elements they contain to be totally ordered (they are implemented with balanced trees) and provide iterators that goes through their elements in order.

Container classes have similar interfaces that allow them to be used interchangably in many cases. E.g., most classes define a size member that returns the number of objects they contain. Inefficient operations, such as random access to a linked list, are not provided, but can be added. Thus, an algorithm that uses random access will not compile with a linked list container, but it will with a vector container. Overall, standard library containers are usually just as efficient as an equivalent customized data structure.

The use of vector is typical:

```
#include <vector>
#include <iostream>
using std::vector;   // Bring the std::vector and
using std::cout;     // std::cout namespaces into scope.

void main()
{
```

```
vector<int> myvec;          // Create an empty vector
myvec.push_back(10);        // Append 10 to the end
myvec.push_back(11);        // Append 11
myvec[1] = 3;               // Change the 11 to 3

// Print the members
for (int i = 0 ; i < myvec.size() ; ++i )
  cout << myvec[i] << '\n';

// Print the members using an iterator
for (vector<int>::const_iterator it = myvec.begin() ;
     it != myvec.end() ; ++it )
  cout << *it << '\n';
}
```

The standard library provides algorithms (search and replace, sorting, heap operations, and a permutation generator) that operate with containers and their iterators.

The `find` algorithm is one of the most basic. It returns an iterator pointing to the element that matches a given value or an iterator pointing just past the end of the container. It takes two iterators that specify the first and one past the last element to be searched along with the value to locate. For example,

```
list<int> l;

list<int>::iterator i = find(l.begin(), l.end(), 10);
if (i != l.end()) {
  // *i == 10 here
}
```

The `for_each` operation applies its third argument using the () operator to each member of the subsequence described by the iterators that are its first and second arguments. Since algorithms, including `for_each`, are templates, the third argument may be a simple function or a template for a class implementing the () operator:

```
void print(int v) { cout << v << ' '; }

void f(vector<int>& v)
{
  for_each(v.begin(), v.end(), print);
}
```

```
template<class T> class Print {
  ostream& s;
public:
  Print(ostream& ss) : s(ss) {}
  void operator()(T& x) { s << x << ' '; }
};

void g(list<int>& l)
{
  Print p(cout);
  for_each(l.begin(), l.end(), p);
}
```

To better support algorithms, the standard library provides a variety of templates that come close to making C++ behave like a functional programming language, providing such operators as bind1st, which returns a unary function that calls a binary function with its first argument fixed.

8.10   History

Bjarne Stroustrup's book *The Design and Evolution of C++* [72] is a unique book that describes why he, the language designer, made the choices he did.

Stroustrup began the design of C++ in 1979 wanting a language to help him simulate a large distributed system. His experience with the Simula language provided much of the impetus. This is an object-oriented language containing many of the C++ concepts, but its implementation is less efficient for a variety of reasons, including dynamic type binding and the absence of stack-based storage (everything is placed in the free store).

BCPL, C's grandparent, was another inspiration. Stroustrup liked the efficiency of the programs it generated, and its support for separate compilation allowed it to be used for large systems. However, its type system is even weaker than C's and it provides few mechanisms for structuring programs.

Stroustrup called his first attempt (ca. 1980) C with Classes. Developed largely for use within the Computer Science Research Center of Bell Labs, the syntax of this language resembled modern C++, but only included classes, single inheritance without virtual functions, constructors and destructors, and public/private access control. Within a year, inline functions, default arguments, and assignment operator overloading were also added.

C with Classes was implemented as a preprocessor that translated its input into C for compilation. Thus, C link compatibility has been around from the beginnings of the language.

C++ was the first to advocate the typed parameter style of function declaration, i.e., using

```
int foo(int a, double b) { /* ... */ }
```

instead of the K&R style

```
int foo( a, b )
int a;
double b;
{ /* ... */ }
```

This feature was proposed first for C++ and was eventually adopted in ANSI C.

Stroustrup considered C with Classes to be a medium success—useful enough to support a developer but not a development group.

Unlike C, which was designed for very small systems by today's measures, Stroustrup assumed one MIPS and one megabyte was available. In 1982, this was a bit optimistic, but is certainly reasonable now. Partially because of this, though, C++ generally compiles much more slowly than the equivalent program written in C.

Stroustrup developed the first real C++ compiler, Cfront, starting in 1982. Like C with Classes, the C++ "compiler" was actually a preprocessor that generated C code to be compiled. This approach naturally lead to longer compilation times, but the portability advantages were worth it. Most modern C++ compilers do not operate as preprocessors to improve compilation speed.

Virtual functions was the most important addition to C++ after C with Classes. Overloading of arithmetic operators came next.

References were added partially to support efficient overloaded operators. Without references, large objects would have to be copied or the code changed to explicitly pass pointers.

8.11   Implementing Inheritance and Virtual Functions

One of the main technical tricks in C++ is implementing inheritance, virtual functions, and multiple inheritance with minimal overhead. Single inheritance simply adds new fields to the end of existing class representations. Virtual functions add a field to each object that points to the table of virtual functions for the object's class. Multiple inheritance adds a base-class offset to each function in a virtual table.

Member functions behave like regular functions with an implicit this argument that points to the object. A call to a non-virtual member function passes the "this" argument:

```
// C++ code                     /* equivalent C code */
class B {                       struct B {
public:                            int field;
  int field;                    };
  void addto(int i);
};

void B::addto(int i)            void B_addto(B* this, int i)
{                               {
  field += i;                     this->field += i;
}                               }

B obj;                          B obj;
obj.addto(10);                  B_addto(&obj, 10);
```

A derived class contains the fields of the base class followed by those in the derived class. When a base class function is called on a derived class, the extra fields are simply ignored.

```
class D : public B {            struct D {
public:                            int field;
  double field2;                   double field2;
};                              };

D obj2;                         D obj2;
obj2.addto(3);                  B_addto((B*)&obj2, 3);
```

Each object of a class with virtual functions includes a field that points to a table of the class's virtual functions.

```
class B {
public:
   virtual int f();
   virtual int g();
   int x;
};
```

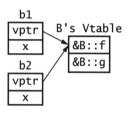

```
class D : public B {
public:
   int f();
   virtual int h();
   int y;
};

B b1, b2;
D d1;
```

Multiple inheritance with virtual functions is a bigger challenge. Consider the following:

```
class A {
public:
   virtual void f();
   int x;
};

class B {
public:
   virtual void f();
   virtual void g();
   int y;
};

class C : public A, public B {
public:
   void f();
   int z;
};

C c1;
B b1, *b2 = &c1;
```

Casting a pointer from a derived class to a base class may require changing it. For example, casting the address of c1 to a pointer to B requires moving it past an A object (a vptr followed by x). As is necessary, this leaves b2 pointing to the beginning of a B object.

When b2->f() is called, it must invoke C::f() with this set to a pointer to a C. How to do this is stored as a correction associated with each function in a virtual table. Thus, calling b2->f() adds f()'s correction (here, -2: one for vptr, one for x) to this before invoking C::f().

Adding a correction to each function in a virtual table works, but adds overhead even when a program uses nothing but single inheritance. An alternative is to use thunks: small pieces of code that adjust arguments before entering the body of a function. This is hard to code in C, but is natural in most assembly languages.

Only one thunk is needed for this example since only one correction is non-zero. The thunk for calling C::f() for a B within a C subtracts 2 from this before jumping to the entry of C::f:

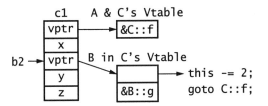

8.12   Exercises

8–1.   How would you achieve the effect of reference arguments in C? Would you have to change how functions with reference arguments are called?

8–2.   What are the advantages of virtual functions? What are the disadvantages? How might you implement them in C? Hint: function pointers.

8–3.   What is the main problem that using namespaces can avoid?

8–4.   Contrast using a template to write a generalized array sorting procedure with using virtual functions. What are the advantages and disadvantages of each? How does the code size and speed compare in each case?

9

Java

Sun's Java programming language is a newer language based on C++, but incompatible with it. Like C++, Java is an object-oriented language that provides classes and inheritance. It is a higher-level language than C++, providing object references, arrays, and strings instead of pointers. An automatic garbage collection facility frees the programmer from most memory management worries.

Unlike C or C++, a Java program is purely a set of classes: there are no standalone functions. Like C++, a new type defines its representation and the operations that may be performed on it.

Java omits many of the more complicated C++ constructs including templates, namespaces, multiple inheritance, and operator overloading. Most were omitted for good reason. The generic container classes in C++'s STL (the main use of C++ templates) are part of the Java language. Java's package mechanism is simpler and more structured than C++'s namespaces. Java's interfaces replace C++'s multiple inheritance with a simpler, more restricted mechanism.

Java is implemented with an interpreter to make its object code portable. This interpreter, the Java Virtual Machine or JVM, is part of the language, so a compiled Java program should run on any machine with a JVM. Unfortunately, JVM and library incompatibilities often prevents this, so Java is more often used to program web servers and other applications where code portability is less crucial.

Java's security features are novel for a general-purpose programming language. Because of Java's design, it is possible to verify that a

Java program will not do anything malicious, such as modify or transmit files. This facility permits World-Wide Web browsers to safely run code from unknown sources.

Slow execution is the main drawback of the interpreted approach. Although this technology is improving, interpreted Java programs only run at about one tenth the speed of a similar compiled C++ program. Reducing this penalty remains an active area of research.

Java continues to evolve. The language itself is fairly stable, but its extensive libraries are in a continuing state of flux. Most programs use these libraries extensively, so this frequently limits portability.

9.1 Types

The sizes of Java's eight primitive data types are precisely defined:

`boolean`	either `true` or `false`
`char`	16-bit Unicode character
`byte`	8-bit signed integer
`short`	16-bit signed integer
`int`	32-bit signed integer
`long`	64-bit signed integer
`float`	single-precision IEEE 754 floating-point (32 bits)
`double`	double-precision IEEE 754 floating-point (64 bits)

Java uses the Unicode international character set [78] for its characters and for its source text. This makes manipulating a language like Japanese as easy as manipulating English, and non-English identifiers are allowed in a program. Unicode characters consume twice as much storage as their ASCII counterparts.

Java uses object references instead of pointers. Any variable of a class type is actually an object reference. Its default value is `null`, meaning it points to no object. New objects are created with the new operator. Assigning to an object reference only changes to which object the reference refers, not the original object. Applying a method to the object through a reference may change the object. All references to the object will also see this change.

Safety is the main advantage of object references. Since a Java program cannot make an object reference point at something that is not an object, Java programs cannot have the most common com-

```
import java.io.*;

/** Print lines from standard in or files matching a pattern. */
public class Mygrep {

  /** Print lines in the stream matching pattern */
  public static void printmatching(Reader is, String pattern)
      throws IOException {
    BufferedReader br = new BufferedReader(is);
    String line;
    while ( (line = br.readLine()) != null )
      if ( line.indexOf(pattern) >= 0 )
        System.out.println(line);
  }

  /** Usage: java mygrep pattern [files...] */
  public static void main(String[] args) {
    if (args.length < 1) {
      System.err.println("Usage: Mygrep pattern [files...]");
      System.exit(1);
    } else {
      String pattern = args[0];
      try {
        if (args.length == 1)
          printmatching(new InputStreamReader(System.in),
                        pattern);
        else
          for (int i = 1 ; i < args.length ; i++) {
            try {
              FileReader f = new FileReader(args[i]);
              printmatching(f, pattern);
              f.close();
            } catch (FileNotFoundException e) {
              System.err.println("Mygrep: can't open "+args[i]);
              System.exit(1);
            }
          }
      } catch (IOException e) {
        System.err.println("Mygrep: I/O error");
        System.exit(1);
      }
    }
  }
}
```

Figure 9.1: A Java program that prints lines that match a pattern.

mon bug in C/C++ programs: dereferencing erroneous pointer values. Dereferencing a null object in Java instead throws an exception that can be caught. Furthermore, Java's automatic garbage collection ensures that an object being referred to will not be deleted.

The Java language provides a string class. Normal strings can only be read, but the StringBuffer class, part of the standard library, supports mutable strings. String expressions use StringBuffers for intermediate string results.

The String class provides methods for concatenating, searching for characters or substrings, comparisons, and conversions to and from primitive types and arrays of bytes and chars. Objects of the String class are constant, but StringBuffer class supports inserting or replacing single characters or substrings and appending to the end of string.

Java provides arrays of objects with the same type. Arrays are objects, so in addition to reading and writing their contents, their length may be queried. Moreover, it is possible to assign an array to an array variable (an object reference). Constructing an array fixes its size; the Vector standard class is an array that may grow or shrink.

Arrays are declared with a pair of brackets following the type name. (This differs from C's declarator syntax where the brackets are part of the name being declared.) An array's size is not part of its type, but is specified when a new array object is created. Thus,

```java
// arr1: an object reference of array type, initially null
int[] arr1;

// arr2: a reference to an array of 20 ints
int[] arr2 = new int[20];
```

Arrays of objects permit polymorphism. If class E extends class B, then the array type E[] extends the array type B[].

Arrays of arrays can be created. Since each is a separate object, each subarray may be a different size, although constructing arrays of the same size is syntactically easy.

```java
double[][] mat1; // Initially null

// A 4x5 array of zeros
double[][] mat2 = new double[4][5];
```

```
// An array of 10 null array references
double[][] mat3 = new double[10][];
```

9.2 Expressions

Java's menagerie of operators essentially matches C's (see Figure 7.2 on Page 121). Notably absent are pointer-related operators such as dereference (*) and structure dereference (->). The field reference operator . is used instead.

Java's unsigned right shift operator >>> always fills with zeros. This is needed because Java does not have C's unsigned types. The signed right shift operator >> fills with the signed bit, making it behave as an integer division by a power of two.

Java expressions are evaluated in the order in which they are written; the compiler is not free to rearrange them for efficiency.

9.3 Classes

All variables and methods in Java are associated with some class. Like a C++ class, a Java class creates a new type by defining data in its representation (fields) and operations to perform on it (methods). Like C++, members (fields or methods) may be declared static to associate them with the class, rather than objects of the class.

Each member of a class is accessible to every other, but its visibility outside the class can be set to one of four policies. By default, a member is only visible to other classes in its package. Declaring it private restricts access to only members of the same class. A protected member can also be accessed by members of inherited classes. Finally, a public member is accessible to all.

Marking a method or class final prevents a derived class from overriding it. The final applied to a method behaves in a way opposite from C++'s virtual. A field marked final is a constant.

The final directive allows optimization. When a method is invoked on an object, the interpreter must examine the object's type to resolve which version of the method (i.e., if it is overridden in derived classes) should be invoked. By marking a method final, the compiler knows no class may override it, and it is usually possible to resolve which implementation to invoke. For small methods, a compiler may be able to inline the method by replacing its invocation with a copy of its body.

Java permits overloaded methods: two or more methods with the same name but different arguments in the same class. When called, the most specific method with a matching number of arguments is invoked, and it is an error if this is not unique. A method is less specific if all its arguments can be assigned to another method's arguments.

It is often useful to define a class not for producing objects but as a common base for other classes. Marking any method in a class `abstract` makes the class abstract in this sense. An attempt to create an object of this class produces an error. A class that extends an abstract class must supply definitions for every abstract method to avoid being an abstract class itself.

All Java object implicitly extend the `Object` class and inherit a variety of general-purpose methods. The `equals` method returns true when two objects have equivalent contents. (By contrast, the `==` operator is true only if two object references point to the same object.) The `hashCode` method returns a hash code for the object that allows it to be quickly located in a `Hashtable` object, part of the standard library. The `getClass` method returns the `Class` object that created the object. The `clone` method creates a new object with a copy of the object's state. Other object methods include `wait` and `notify`, used for thread control and described in Section 9.6.

A `static` block defined in a class runs when the program starts and usually used to initialize the class's static members. The order in which different classes' blocks run at the start of the program is undefined, so other classes' static fields may have their default values of 0 or `null`.

A class may contain `native` methods that are implemented in another language. At runtime, the Java program calls the native method, which must be linked with the runtime environment.

9.4   Interfaces

With normal class inheritance, a derived class inherits both the interface of the base class (the methods, their arguments, etc.) and the

implementation of those methods. Java's interface mechanism provides a way to inherit just an interface without the implementation.
Specifically, an interface is a class containing method and constant
declarations that a class may choose to implement, meaning it con-
forms to the interface and implements its methods.

```java
import java.io.*;

interface Printable {
  void print(PrintStream out);
  void setIndent(int indent);
}

class PrintMe implements Printable {
  private int ind = 0;
  public void setIndent(int indent) { ind = indent; }

  public void print(PrintStream out) {
      for (int i = 0 ; i < ind ; i++) out.print(' ');
      out.println("Hello World!");
  }

  public static void main(String[] args) {
    PrintMe p = new PrintMe();
    p.setIndent(10);
    p.print(System.out);
  }
}
```

Unlike normal (single) inheritance in Java, a class may choose to
implement two or more interfaces. Thus, interfaces provide part of
the functionality of C++'s multiple inheritance. Java's designers omitted multiple implementation inheritance because of the confusion it
can cause when there are common base classes. (C++'s virtual base
classes address this problem, but it remains confusing.)

9.5   Exceptions

Java provides an exception throwing mechanism for error handling.
When an error occurs, a method may throw an Exception object.
The method terminates and the call stack unwinds until it reaches a
try block that catches that exception type.

9.6 Threads

The Java language provides asynchronous threads that communicate through shared memory. It provides facilities for communicating between threads, waiting for a thread to terminate, grouping threads for protection, and preventing contention over objects.

A `Thread` object represents a thread. Its `start` method starts the thread running, and its `join` method waits until it terminates or optionally times out. The thread's body is supplied by either extending `Thread` and overriding its `run` method, or passing a `Runnable` object (i.e., one with a `run` method) to `Thread`'s constructor.

Each thread has a priority providing rough control over which threads may interrupt it. Higher-priority threads usually get preference, but the mechanism is imprecise and unreliable.

Although writing an integer or smaller is guaranteed to be atomic, writing a non-`volatile double` or `long` may not be. This becomes an issue when multiple threads are attempting to read or write the same value at the same time. If the value in question is a single integer, there is no danger of interleaved execution. However, if the value is a `long`, or multiple values within an object, this can cause problems.

Java provides a mechanism to prevent a thread from reading an object's partially-updated state while another thread is writing it. Each object has a lock designating which thread currently owns the object. A block of `synchronized` code waits to acquire its object's lock before running, and it releases the lock when the block terminates.

The `synchronized` construct is not a panacea since it only prevents other `synchronized` blocks from running. The programmer is responsible for enclosing every block that might cause such conflicts with `synchronized`. Java's standard libraries are written this way.

To allow threads to react to events, a thread can suspend itself with a call to `wait` and wait for an object to reawaken it with a call to `notify` or `notifyAll`. This can be used for producer-consumer relationships. A consumer method synchronizes on its object and repeatedly calls its thread's `wait` method until data is available. Each call to `wait` releases the threads's locks, suspends the thread, and adds it to the object's list of waiting threads. A producer method synchronizes on its object, writes new data, and calls `notifyAll` to awaken all threads waiting on the object for the new data.

Abstractly,

```
class XmitRcvr {
  public synchronized void transmit(Data d) {
    // ... write data into object ...
    notifyAll();
  }

  public synchronized Data receive() {
    try {
      while (/* data not available */)
        wait();
    } catch (InterruptedException e) {
        return null;
    }
    /// ... return the data ...
  }
}
```

Here is code for a one-place buffer.

```
class OnePlace {
  Element value;

  public synchronized void write(Element e) {
    while (value != null) // Block while full
      wait();
    value = e;
    notifyAll();  // Awaken any waiting read
  }

  public synchronized Element read() {
    while (value == null) // Block while empty
      wait();
    Element e = value;
    value = null;
    notifyAll();  // Awaken any waiting write
    return e;
  }
}
```

The sleep method allows a thread to suspend itself to be reawakened a specified time in the future. The delay is more a request than a command. The thread will not be awakened earlier, but may be awakened later.

9.7   Packages

Collections of Java classes can be arranged into packages, making it easy to avoid name collisions when assembling programs from pieces. Once a collection of classes is grouped into a packages, it may be used in one of two ways. A single package member can be referred to by placing the package name before the member name. Alternately, some or all package members can be imported, making it possible to access them without using the package prefix. Java's standard libraries are arranged into packages like this.

By default, a package member is not visible outside the package unless it is declared public. This allows the interfaces to the package to be separated from implementation-specific details.

Packages can be arranged hierarchically, since package members may be packages.

9.8   Type wrappers

Every primitive type has a corresponding class that has a few uses. Static members of these classes hold values and members that manipulate the types, such as java.lang.Float.intValue. An object of the class has the value of the primitive type, but since it is an object, it can be used where an object reference is needed. For example, a hash table of integers needs to contain integer objects since the hash table class stores Objects.

9.9   Garbage Collection

The Java language specification does not specify a garbage collection policy, leaving it as an implementation choice. While this provides flexibility, it adds uncertainty to a particular program's behavior.

Garbage collectors are conservative, meaning they are careful to never free memory that could still be in use, but are not guaranteed to free memory that is not still in use. The memory leaked by a conservative garbage collector generally is not a problem for a program running on workstation, since such programs often terminate quickly and have access to vast amounts of virtual memory. But for an embedded system, whose programs must run indefinitely in a cramped memory space, the memory leaked by a conservative garbage collector can be fatal.

There are two main approaches to garbage collection. The stop-and-copy approach pauses the program, copies all reachable memory from the running memory space to a new one, and releases the old memory space. The mark-and-sweep approach marks all reachable memory and releases all bits of memory that was not marked (i.e., sweeps it away).

All garbage collection algorithms must distinguish live objects—reachable from the root set of live variables—from dead ones. Doing this is exactly is difficult because pointers must be distinguished from non-pointers. Although a Java compiler knows this information, a garbage collector often does not for efficiency reasons. Instead, garbage collectors are often conservative, assuming everything that can be a pointer is. In the process, integer, floating-point, and character data may accidentally interpreted as pointers. This causes the garbage collector to preserve more data than it should, but it will never accidentally discard data.

9.10   Java in Embedded Systems

Sun's standard Java implementation is a sprawling twenty-megabyte beast best suited for enterprise applications such as generating web pages from relational databases. Memory and performance are less critical in such environments, but embedded systems generally cannot afford a memory footprint of this size.

Java's size presents a problem for embedded systems that is still being solved as this book is being written. Developers have been demanding scaled-down versions of Java, and a number of companies have been developing such environments, but few standards are in place and there has been as much posturing on the part of large companies as there has been serious technical progress.

I am confident such problems will be resolved in the years to come. Java is too good a language and the momentum behind it is too great for these problems to remain unsolved. Nevertheless, it will be a while before Java is as perfectly suited for use in embedded systems as C.

Device drivers—routines that talk directly to the hardware—are a perpetual challenge in embedded systems. Such drivers need to be efficient and interface directly with the lowest-level I/O, things Java is not well-suited for providing. For example, device drivers generally need direct access to memory to access memory-mapped I/O, yet Java specifically prohibits direct access to memory for safety reasons. Mechanisms for direct memory access for Java have been proposed, but none are standard.

Most interesting embedded applications employ interrupts: asynchronous signals from the environment that request a processor stop what it is doing and examine the source of the interrupt. Communication channels and sensors are typical sources of interrupts. Java does not provide a standard mechanism for creating programs that can respond to interrupts (in C, a function can be registered as an interrupt service routine), but it needs such a mechanism to find wide use in embedded systems.

For traditional multitasking applications where there are no hard real-time deadlines, Java's somewhat vague specification of its concurrent scheduler is good enough. But many embedded systems must respond to hard realtime deadlines. Priority-based preemptive schedulers are usually the solution, but Java's scheduler guarantees little. This is yet another issue that the real-time Java community is attempting to address.

The nondeterministic garbage collection systems in many Java environments are also a problem for hard real-time systems. Meeting hard deadlines demands predictability, so occasionally freezing for an unpredictable amount of time to collect garbage is unacceptable. Garbage collection does not have to behave like this, and proposals have been made to provide more control over garbage collection behavior for embedded applications.

9.11   Exercises

9–1.   Why does Java define widths of basic types (C lets them vary)?

9–2.   Why does Java not have need for C++'s templates?

9–3.   List three things currently missing from Java that are needed by most embedded software systems.

10

Operating Systems

An operating system (os) is a program that provides an environment for executing other programs, often providing facilities for I/O, a filesystem, networking, virtual memory, and multitasking: a way to run multiple programs concurrently on a single processor.

Simple embedded systems do not need an operating system. They run a single program that communicates directly with their peripherals at a one processing rate. Examples include digital watches, microwave ovens, and calculators. These systems are often implemented with a cyclic executive: a single loop such as in Figure 10.1.

When inputs cannot wait until the next iteration of a cyclic executive, interrupts are the solution. When a source of data raises an interrupt, the processor briefly switches to an interrupt handling routine that acknowledges the interrupt and copies the new data into memory before resuming the original task. Figure 10.2 illustrates this.

This approach has the limitation that everything happens at the same rate. Although interrupts are serviced quickly, data read at the interrupt is processed at the rate of the inner loop regardless of how

```
for (;;) {
  while (!inputs_ready())
    ; /* wait */
  read_inputs();
  process();
  write_outputs();
}
```

Figure 10.1: A simple system using a cyclic executive. In each iteration, it polls its inputs before processing the information. This fails when new data demands immediate attention.

```
void main()
{
  register(int_handler);
  for (;;) {
    while (buffer_empty())
      ; /* wait */
    read_from_buffer();
    process();
  }
}

void int_handler()
{
  Data d = read_data();
  acknowledge_interrupt();
  write_data_to_buffer(d);
}
```

Figure 10.2: A simple interrupt-driven system. It registers an interrupt handling routine before entering into a loop that polls a buffer before processing its contents. The interrupt handler copies data into the buffer before acknowledging the interrupt.

quickly it is needed. The fundamental problem is that operations in the inner loop were scheduled and fixed when the system was designed, leaving no flexibility when data arrives at unexpected times.

Preemptive multitasking is the solution when data arrives at unpredictable times and demands immediate attention. Instead of a single inner loop, a system is described as a collection of independently-running threads of control—processes. The operating system runs these processes based on a scheduling policy, and can switch between them as it deems necessary to meet deadlines.

Splitting a system into processes separates functionality from timing. Unlike in a cyclic executive, where both the functions and the order in which they are performed are controlled by the layout of the main loop, a process contains only its function; the operating system's scheduler determines what functions are performed when.

Modularity is another advantage of processes. Since processes are separated and often define their inputs and outputs, the system is easier to understand and modify. Some oss also prevent a process from writing into another's address space, making it very difficult for an errant process to damage the whole system.

The scheduling policies of the two classes of multitasking operating systems have different objectives. Most computers run a timesharing os whose main goal is fairness: providing similar interactive response times for all running programs. A typical timesharing os periodically switches between each process, perhaps once every tenth of a second. By contrast, most embedded systems run a real-time operating system (RTOS) whose scheduling objective is to meet deadlines, i.e., to make sure each process that must complete by a certain time does so. An RTOS usually runs a process to completion unless it is preempted by a higher-priority process that needs to complete sooner.

This chapter focuses on the concurrency model provided by multitasking operating systems in general and RTOSes in particular, since this can greatly affect the behavior of concurrently-running software. Understanding and controlling concurrency is often one of the most crucial aspects of real-time embedded system design.

10.1   Timesharing Systems

In a timesharing system, the scheduler's goal is to provide all processes with acceptable interactive performance. Fairness is the objective: no one user should be able to monopolize the processor's resources at the expense of others.

Timesharing systems are ideal for delay-tolerant batch or interactive systems. Batch systems, such as those that print your bank statement, must run and terminate in a reasonable amount of time, but when they produce results is not important. Interactive systems such as graphical user interfaces interact with humans, who tolerate small delays and occasionally big ones.

A timesharing system's scheduler usually timeslices: divides time into a series of equal periods and assigns each running process a slice in a round-robin order. This approach balances average response time at the expense of many context switches.

The processes in a timesharing system often have a priority number. Lower priority processes are generally intended to be long jobs running "in the background" where predictable interactive performance is not important.

Many textbooks discuss timesharing operating systems. Both Silberschatz and Galvin's "Dinosaur book" [71] (from the cover illustration) and Tanenbaum [75] are popular. Leffler et al. [53] and McKusick et al. [58] describe the design of the BSD variant of Unix.

10.2 Real-Time Operating Systems

In a real-time environment, the time at which computation is done can be more important than the result. Small errors in the output of a car's brake controller might make for a jerky ride, but an output that comes two seconds late could kill the driver.

Hard real-time systems live in the world of power, speed, and steel, where missing a deadline can be catastrophic. You find these systems controlling nuclear power plants, aircraft control surfaces, and automobile engines. By contrast, soft real-time systems may miss deadlines. In this category are cell phones, DVD players, and other systems that mostly interact with humans, who get annoyed when the system misses a deadline, but rarely explode.

Consider a software digital answering machine. It might consist of a process responsible for playing messages and another that handles the user interface. A periodic clock based on the sampling rate triggers the playback process, which must write the next sample before the next clock for acceptable sound quality. The user interface process, by contrast, has less stringent deadlines. An RTOS would ensure the playback process runs frequently enough to meet its deadline and run the user-interface process in the time remaining.

Real-time operating systems try to meet deadlines using a simple principle. Just as you would switch to a more important task due sooner if your boss gave one to you, an RTOS suspends low-priority processes when one with higher priority starts.

A typical RTOS uses a fixed-priority preemptive scheduler. In this approach, each process has a priority number that may be assigned before the system runs. The RTOS always runs an active process with the highest priority. Starting a higher-priority process preempts any lower-priority process, but the scheduler does not timeslice. Instead, a process generally runs to completion unless it is preempted. If another process with the same priority starts, the currently-running process is allowed to complete before the new process is started.

Every RTOS supports interrupts. When an RTOS receives an interrupt, it briefly switches to the interrupt service routine for that interrupt. This routine is usually very short and simply copies new input data into a buffer, acknowledges the interrupt, asks the RTOS to start a process that handles the newly-arrived data, and terminates. The idea is to respond as quickly as possible to the interrupt (this time is called interrupt latency) without upsetting running processes. Such an approach separates the arrival of data from the need to process it. If the process started by the interrupt routine is at a higher priority than the currently-running process, the RTOS will switch to the new process, otherwise the new data will sit in a buffer until the RTOS can run the process to handle the data.

Predictability and speed are the main criteria for judging an RTOS. The time an RTOS takes to do anything needs to be predictable, be it the time to switch between processes, service an interrupt, or allocate and release memory. Unpredictability can make it impossible to guarantee response times—unacceptable in hard real-time applications. Naturally, the second most important criteria is how quickly an RTOS can do any of these things. An interrupt can be lost if it takes too long to service. Spending too much time performing context switches wastes processing power and can cause deadlines to be missed.

The size of an RTOS is also important. Since memory in an embedded system is invariably cramped, a smaller RTOS is better. RTOS designers have addressed the size problem through careful coding and modularity. A typical RTOS is divided into modules that can be included or excluded depending on their need. An embedded system with no need for networking does not want to waste space for its code.

10.3   Real-Time Scheduling

The traditional model for real-time task scheduling (Figure 10.3) consists of a collection of tasks, each with an initiation time, a deadline, an execution time, and a period. The assumption is that every task is periodic with a known period, but that the phases are unknown, an assumption that accounts for unrelated periods and jitter. The deadline for a task is generally the end of its period.

Sporadic tasks are modeled by assuming a minimum initiation interval, i.e., by assuming they are periodic with a minimum period. Tasks that start at truly random times are essentially impossible to handle, since an unlimited number of them could start within a vanishingly small period of time.

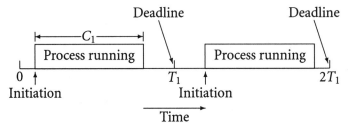

Figure 10.3: The model of a real-time process. It is executed regularly with period T_1. C_1 is its computation time (assuming no other process is running). It may be initiated any time during its period. Its deadline is typically the end of the period.

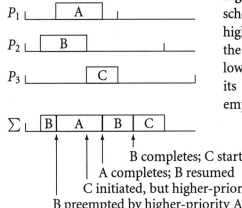

Figure 10.4: Rate-monotonic scheduling. Process P_1 has the highest priority since it has the shortest period; P_3 has the lowest. A would have missed its deadline had it not preempted B.

B completes; C started
A completes; B resumed
C initiated, but higher-priority A continues
B preempted by higher-priority A

10.4   Rate-Monotonic Scheduling

In 1973, Liu and Layland [55] defined what has become the standard approach: rate-monotonic scheduling. It assumes a single processor system with fixed-priority tasks, no communication, and a preemptive scheduler. To meet its goal of meeting all deadlines for any valid combination of initiation times, it assigns the highest priority to the task with the shortest period, the next lowest priority for the next shortest period, and so on. It ignores computation times. Within its fixed-priority framework, rate-monotonic scheduling is optimal.

Figure 10.4 shows how a system with three processes might behave with rate-monotonic scheduling.

The rate-monotonic approach is based on the idea of a critical instant—when all tasks are initiated simultaneously. Intuitively, the shortest-period task should run first, since it needs to finish before any tasks with longer periods.

Briand and Roy [15] describe this approach in greater detail.

The rate monotonic approach suffers from ignoring the dynamic behavior of the system, and is instead designed for the worst-case. Such an approach often leaves the processor underutilized.

Earliest deadline first (EDF) is an alternative scheduling technique that changes tasks' priorities on the fly so that tasks with earlier deadlines get higher priorities. This can achieve higher utilization.

10.5   Priority Inversion

Priority inversion, which occurs when a lower-priority task acquires a resource and is preempted by a higher-priority task that wants it, can cause a deadlock. The scheduler will suspend the lower-priority task in favor of the higher-priority one, but the higher-priority task will block waiting for access to the resource since the lower-priority has the resource and is itself blocked by the higher-priority task.

The solution, called priority inheritance, is to raise the priority of the lower-priority task while it has the resource and lower the priority when the task releases the resource. Most RTOSes provide facilities for priority inheritance.

10.6   Interprocess Communication

Processes frequently need to communicate, so most RTOSs provide a variety of mechanisms for doing so.

Shared memory—where two or more processes can read or write the same section of memory—is the most basic mechanism in software. In fact, processes always share memory unless an OS separates them using virtual memory.

Processes communicate through shared memory by agreeing to read and write certain locations. This works provided processes access the data sequentially (a process finishes writing an object before another attempts to read or write it). The danger comes when another process attempts to read or write partially-written data. Either causes a hard-to-find bug when the processes read back gibberish.

Semaphores or locks (Java uses these—see the `synchronized` attribute in Section 9.6), are the usual solution to this problem. By convention, processes agree not to read or write a particular section of memory unless they have acquired a lock for that area. Locks are built such that they can be owned by at most one process at any given time, so processes that use idioms such as

```
lock 1;
while ( !acquire_lock(1) )
  /* wait */;
... read or write data associated with the lock
release_lock(1);
```

are guaranteed safe access to data associated with the lock.

Semaphores are a basic synchronization mechanism that can be used to build more complicated types of communication. For example, a mailbox, a shared storage location that may contain a message, can be built using semaphores. Messages are atomic so mailboxes do not suffer from the same problems as raw shared memory.

A queue is a buffered communication scheme that can also be built with semaphores. A queue maintains the order of messages and can contain more than one. If one process writes a sequence into a queue, another process can read the same sequence from the queue, and these two operations do not need to be synchronized. The queue can hold as much or as little of the sequence as needed based on the relative execution rates of the two processes.

10.7 Other Services

Operating systems often provide hardware abstraction, that is, high-level access to peripherals that allow a program running under the operating system to, say, send a packet to a network address. The operating system performs such a request by adding packet header information, deciding on a route, and ultimately toggling bits in the control registers for the network interface. This last part is usually done by a piece of software called a device driver. Well-integrated device drivers are often the most useful aspects of any operating system.

Operating systems often provide security by denying programs access to certain resources, such as peripherals (device drivers should take care of this) or the memory spaces of other programs. Security can prevent erratic or rogue programs from crashing a system—an especially critical feature in most high-reliability systems.

Since real-time operating systems are often used in embedded systems, they often provide elaborate debugging facilities. Unlike normal software development, where the development environment and the system are often the same computer, embedded systems are not usually capable of being their own development environment. For debugging, this creates a problem because the debugger (the program able to observe system operation) usually runs on a separate computer and must communicate with the embedded system. Many different communication mechanisms exist, often highly dependent on the hardware available, and many commercial rtoses tout their ability to handle many of them.

Many operating systems, especially those designed for desktop or larger machines, provide a filesystem—a mechanism for storing and retrieving data from mass storage, typically hard disk drives. Issues with filesystems include buffering and caching (it is faster to read a few kilobytes of data from disk and store it in memory than reading a single byte from the disk at each request) and managing fragmentation (Deleting a file can leave a "hole" in the storage. Using this hole can cause the storage for a file to be non-contiguous).

More and more operating systems supply a set of routines for managing a graphical user interface. Microsoft Windows and the Apple Macintosh os are two familiar desktop os examples, but more and more embedded systems are requiring such facilities. Size is usually the main issue with such libraries. A comprehensive set of facilities can consume hundreds of megabytes, making it impractical for most smaller embedded systems.

Device drivers and protocols for networking are now part of most embedded systems. Such software is often called a stack because it is built in layers: the lowest layer talks directly to the network interface hardware. The layer above that understands the lowest hardware protocol (e.g., Ethernet), the layer above that speaks the next-highest protocol (e.g., the Internet protocol IP). At the top are high-level protocols such as HTTP on which applications are based.

After correctness, efficiency is the main issue with networking software. A simpleminded implementation of a network stack copies incoming and outgoing data much more than needed (caused by transferring data between protocols); more clever implementations avoid this as much as possible.

10.8   Exercises

10–1. How does splitting a program into processes help in meeting deadlines?

10–2. List the differences between a timesharing OS and an RTOS.

10–3. What are the differences between batch, interactive, and real-time systems?

10–4. What is priority inversion? Why is it necessary?

10–5. What are the differences between rate-monotonic and earliest deadline first scheduling? Would the behavior in Figure 10.4 change under earliest deadline first?

10–6. What are the advantages and disadvantages of shared memory? How do semaphores help to avoid some of the disadvantages?

Part III

Dataflow

Dataflow languages describe systems of procedural processes that run concurrently and communicate through queues. Although clumsy for general applications, dataflow languages are a perfect fit for signal-processing algorithms, which use vast quantities of arithmetic derived from linear system theory to decode, compress, or filter data streams that represent periodic samples of continuously-changing values such as sound or video. Dataflow semantics are natural for expressing the block diagrams typically used to describe signal-processing algorithms, and their regularity makes dataflow implementations very efficient because otherwise costly runtime scheduling decisions can be made at compile time, even in systems containing multiple sampling rates.

11

Kahn Process Networks

In 1974, Gilles Kahn wrote a short paper [44] describing a simple language for parallel processing that provides a theoretical basis for dataflow computation. A system in Kahn's language is a set of sequential processes running concurrently that communicate through single-sender, single-receiver FIFO queues. A process that tries to read from an empty queue waits until data is available and cannot ask whether data is available before reading. Kahn showed these restrictions make these systems deterministic, that is, the sequence of messages that pass through each queue does not depend on the speed of the processes or the order in which they execute.

Kahn process networks are important for dataflow computation theory because of their deterministic concurrency, but their scheduling overhead makes them impractical. Because it can do nearly anything when it is running, a Kahn network demands a flexible scheduler. One solution is to restrict these networks to allow more compile-time analysis and permit schedulers with less overhead. This is discussed in the next chapter.

11.1   The Language

Figure 11.1 shows Kahn's original example written in a C-like dialect. It defines three processes and five channels, and starts four processes that communicate through the channels. The lower right corner of Figure 11.1 shows the structure of these processes.

```
/* Alternately copy u and v to w, printing each */
process f(in int u, in int v, out int w)
{
  int i; bool b = true;
  for (;;) {
    i = b ? wait(u) : wait(w);
    printf("%i\n", i);
    send(i, w);
    b = !b;
  }
}

/* Alternately copy u to v and w */
process g(in int u, out int v, out int w)
{
  int i; bool b = true;
  for(;;) {
    i = wait(u);
    if (b) send(i, v); else send(i, w);
    b = !b;
  }
}

/* Emit an initial token. Copy u to v */
process h(in int u, out int v, int init)
{
  int i;
  send(v, init);
  for(;;) {
    i = wait(u);
    send(i, v);
  }
}

channel int X, Y, Z, T1, T2;

f(Y, Z, X);
g(X, T1, T2);
h(T1, Y, 0);
h(T2, Z, 1);
```

Figure 11.1: Kahn's example program [44] expressed in a C dialect. (The original was in an Algol dialect.) This starts and connects the four processes as shown in the diagram.

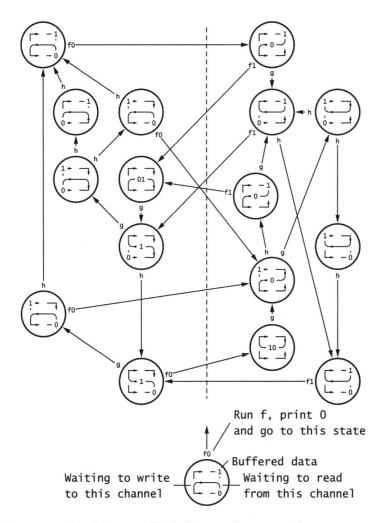

Figure 11.2: Possible runs of Kahn's example system. The system starts in the state in the upper left corner. The system is deterministic because all arcs that print a 0 cross the dashed line from left to right and all that print a 1 cross from right to left.

A process definition resembles a C function definition. The body consists of sequential statements and control is always at a one place in a running process. Arguments of a process include values passed when the process is started (behaving like a C function call) and communication ports. Processes communicate through channels using two functions: wait() waits for and returns the next data value on a channel; send() function writes a value on a channel.

Consider the behavior of the program in Figure 11.1. The two h processes start by sending the value of their init parameters, 0 and 1. The f process copies tokens from alternating inputs to its output, printing each token. The g process alternates between copying a token to its first and second outputs.

The finite-state diagram in Figure 11.2 describes every possible behavior of this fairly simple system. Each state shows the data on each communication channel (zeros and ones). The arcs on the left and right of each state indicate which channels processes f and g will write and read next. The system begins in the state in the upper left, where the two h processes have just sent a 0 and 1 on the two channels leading to f. In the first state, only f can run. This consumes the 0 and copies it to the channel leading to g, producing the top right state. Here, however, two processes can run: f and g. Thus, the scheduler has a choice here. Such choices are what makes scheduling difficult and the state diagram complex.

Figure 11.2 shows the system prints an alternating sequence of o's and 1's regardless of the choices made by the scheduler. The dashed line divides the states into two groups. Every arc along which the system prints a o crosses this line from left to right, and every arc printing a 1 crosses from right to left. Thus, any time the system prints a o, it will print 1 next and conversely.

11.2 Determinism

It is not an accident that system in Figure 11.1 behaves the same regardless of decisions made by the scheduler: all Kahn systems are deterministic in this sense because they restrict interprocess communication to FIFO queues with blocking read semantics. The proof of determinism follows from noting that the sequence of data values each process writes on its output ports (i.e., its behavior) is only a function of the sequence of data values arriving on its input ports. Specifically, the state of each process is only affected by the sequence of values it reads, and not on their arrival time.

Think of a process as executing in two alternating phases. The process runs in the first phase, observing and changing its local variables and writing values onto output queues. Other processes have no affect during this phase. E.g., no other process can read or write variables local to the running process; no variable is shared. So the state of the process at the beginning of the phase completely determines what data values the processes writes during the phase and the state of the process at the end of the phase.

A process reads a data value in the second phase. The data may or may not be available when the process reaches the read, but the process cannot tell if it will have to wait or has waited for data.

Since each FIFO is read and written by exactly two processes, the speeds of the two processes does not affect the sequence of data values sent through the FIFO. This plus knowing the state of each process is only affected by these sequences shows Kahn's systems deterministic.

11.3 Execution

Running a Kahn Process Network correctly is easy; running one without doing more work or using more memory than needed is more challenging. Even the behavior of the small process in Figure 11.2 has many choices that could affect memory consumption. For example, running f in the state in the upper right corner requires storage for two tokens in the center buffer. If the scheduler fires g and then f from this state, the center buffer only needs space for a single token.

A Kahn process network requires dynamic scheduling because its processes can communicate freely when the system is running. Balancing the number of data values produced and consumed is the main challenge.

Two traditional approaches to dynamic dataflow scheduling can cause unnecessary build-ups of data tokens. Data-driven scheduling runs all processes that have enough data available. Unfortunately, this policy can produce tokens faster than they are consumed. The system

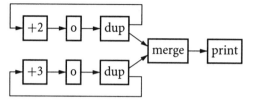

Figure 11.3: A system that fails under data-driven scheduling (after Parks [65, p. 36]). This prints the increasing sequence of integers that are divisible by two or three (i.e., 0, 2, 3, 4, 6, 8, 9, 10,...) by merging two streams that are the multiples of two and three. The two loops produce these streams; the merge node performs an ordered merge of two increasing sequences, discarding duplicates.

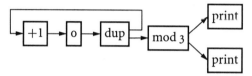

Figure 11.4: A system that fails under demand-driven scheduling (after Parks [65, p. 42]). One print statement prints the integers divisible by three, the other prints all the other integers. The loop generates the sequence 0, 1, 2,... The process labeled "mod 3" copies integers divisible by three to one output and all other integers to the other output.

in Figure 11.3 has this problem. The processes in both loops run constantly under a data-driven policy, but nothing regulates the relative rates of the two loops, whose outputs are consumed at different rates by the merge node. Thus, tokens will accumulate indefinitely on one of the two queues leading into the merge node.

Demand-driven scheduling takes an opposite approach but suffers from the same problems. A process waiting for a token on one of its inputs prompts the scheduler to run the process that could generate the token. This policy can run the system in Figure 11.3, but fails for the system in Figure 11.4. This latter system fails because both print statements demand tokens at the same rate, yet the "mod 3" block produces them at different rates, causing tokens to accumulate without bound on one of the queues leading to a printing process.

In his thesis [65], Tom Parks solved the bounded scheduling problem for Kahn Process Networks by providing a scheduler that executes a system with bounded buffers if it is possible. The idea is simple: begin with a small bound on the size of each buffer. If the program deadlocks because of buffer overflow, increase the capacity of the smallest full buffer and continue. In effect, the scheduler dynamically computes the bounds on the buffers, increasing its estimate only when it proves to be too low.

11.4 Exercises

11–1. Would adding a command `empty()` that reported whether a channel had waiting tokens fundamentally change Kahn Process Networks?

11–2. What is the main disadvantage of running a Kahn Process Network on a standard processor?

11–3. The diagram in Figure 11.2 has a finite number of states. Is this true for every Kahn Process Network?

11–4. Tom Parks' scheduling algorithm always executes a network in finite memory if it is possible. Does it ever use more memory than necessary? Why or why not?

12

Synchronous Dataflow

Synchronous Dataflow (SDF) [52] is a dataflow language in which each process (called an actor) produces and consumes a fixed number of tokens per firing. This behavior makes an SDF system's communication patterns independent of data values, allowing systems to be analyzed completely when they are compiled. Although this also limits what the language can describe, SDF is capable of describing most signal-processing algorithms, even those containing multiple sampling rates. SDF's compile-time scheduling and expressiveness make it a natural choice for many signal-processing algorithms.

Like a Kahn process, an SDF actor (Figure 12.1) has a fixed collection of input and output ports. Each port on an SDF actor is marked with the number of tokens it produces or consumes when the actor is fired. A single-rate actor such as an adder consumes one token from each of its input ports, add the token values, and produce one token on its output port. A 4× downsampling actor consumes four tokens on its input and produce a single token.

For an actor to fire, it must have at least as many tokens at its ports as it will consume. Karp and Miller's Computation Graphs [46] re-

Figure 12.1: An SDF actor with three input and two output ports. Labels indicate the number of tokens consumed or produced when the actor fires.

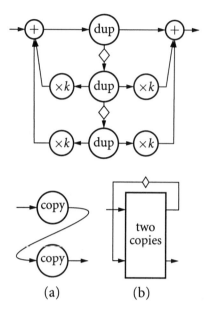

Figure 12.2: A unit-rate SDF system: a second-order IIR filter. The dup actors make two copies of their input. The ×k actor multiplies by a constant. Each diamond indicates an initial token, which behaves like a delay.

Figure 12.3: An example of non-compositionality. (a) A system that copies its input to its output. (b) The two actors have been combined, but an additional delay (the diamond) is needed to avoid a deadlock.

(a)   (b)

semble SDF except they allow the number of token required on an input to exceed the number an actor will consume when it fires. This can be useful for sliding-window filters.

Any useful SDF system must be able to run forever. To do this, it must be possible to fire all actors in a sequence that returns the number of tokens in each communication buffer to its original value. Repeating this firing sequence avoids avoid buffer underflows or the unbounded accumulation of tokens in buffers.

Once this repeating sequence—the schedule—is known, generating code that implements the system is easy. The easiest way is to define each actor as a macro or function and simply string together a call to each actor in the scheduled order. This is particularly attractive for generating code for a DSP since the code for larger, more complicated blocks can be written by hand in assembly.

An SDF system often starts with initial tokens on its arcs to avoid an initial deadlock if the system contains communication cycles. Initial tokens are used because SDF actors are not allowed to emit any initializing tokens, unlike Kahn processes.

Because SDF forces its actors to fire atomically, grouping actors can cause deadlocks. For example, consider the pair of actors in Fig-

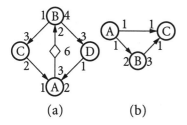

Figure 12.4: SDF systems with (a) consistent and inconsistent (b) rates.

(a)   (b)

ure 12.3a. Each simply copies its input to its output one token at a time. Now consider grouping these two actors as in Figure 12.3b. The system will deadlock immediately unless an initial token is placed on the arc connecting the output to another input. Such behavior can place limits on the largest SDF system that can be built because large systems must be analyzed as a whole instead of in pieces.

Lee and Messerschmitt developed SDF at the University of California, Berkeley [52] starting in the late 1980s. It was the central language for the Gabriel system and later the Ptolemy system [18]. Ptolemy provides additional models of computation that can be mixed.

12.1   Scheduling

The goal in scheduling an SDF system is to determine a actor firing sequence that, when repeated indefinitely, does not accumulate tokens or fire an actor without enough data. Lee and Messerschmitt's procedure [51] first calculates how many times each actor must fire per iteration by solving the rate constraints that the communication channels impose between actors. Next, the system is simulated for an iteration by firing each actor when it has enough data. The technique produces a valid schedule if one exists.

Ignoring the problems of deadlock, the total number of tokens produced on an arc must equal the number of tokens consumed on that arc every schedule period for the system to be consistent. Initial tokens do not change this. Consider the arc in the middle of Figure 12.4a. If a and b are the number of times actors A and B fire each cycle, then for tokens to be produced and consumed at the same rate along this arc, we must have

$$3a - 2b = 0.$$

Repeating this for each arc gives a system of linear equations called the balance equations, usually expressed in matrix notation. For the system in Figure 12.4a, these are

$$
\begin{bmatrix}
3 & -2 & 0 & 0 \\
0 & 1 & -3 & 0 \\
0 & 4 & 0 & -3 \\
-1 & 0 & 2 & 0 \\
-2 & 0 & 0 & 1
\end{bmatrix}
\begin{bmatrix}
a \\ b \\ c \\ d
\end{bmatrix}
= Tp = 0.
$$

Each column of T corresponds to a actor, each row to an arc.

One solution to this is $p = 0$, corresponding to firing no actors, but this is not an interesting solution. If this is the only solution, then the system is inconsistent and no schedule will allow the system to run indefinitely without tokens accumulating. For example, the system in Figure 12.4b has balance equations

$$
\begin{bmatrix}
1 & -2 & 0 \\
1 & 0 & -1 \\
0 & 3 & -1
\end{bmatrix}
\begin{bmatrix}
a \\ b \\ c
\end{bmatrix}
= 0.
$$

The second row implies $a = c$, but adding the first and third rows implies $a + b - a = 0$, so $b = 0$. This plus the second row implies $a = c = 0$.

Mismatched rates causes the problem in Figure 12.4b. The top arc forces A and C to operate at the same rate, but B forces them to operate a ratio of 3:2. This is contradictory, so the system can only do nothing.

Lee and Messerschmitt showed [51] that the balance equations of a connected, consistent system have a family of solutions $p = \alpha q$, where q is a non-zero vector and α is any real number. Intuitively, α corresponds to the number of iterations of the schedule, which, as expected, can be arbitrary. If the solution space has two or more dimensions (i.e., has solutions of the form $\alpha q + \beta r + \cdots$), the system is unconnected and the two or more unconnected pieces should be run separately.

For scheduling, we want a p consisting of small integers. Since the coefficients of the balance equations are integers (an SDF requirement), there is always a rational solution. Multiplying this solution through by its least common denominator forces it to be integer.

For the system in Figure 12.4a, the smallest positive integer solution to the balance equations is $a = 2$, $b = 3$, $c = 1$, and $d = 4$.

Solving the balance equations gives the number of times each actor must fire per cycle to balance token production and consumption rates, but a solution does not always imply the system has a valid schedule. The system may still deadlock: come to a point where no actor has enough input data to fire. The system in Figure 12.4a would deadlock immediately if it had no initial tokens: each actor would be waiting for another to feed it data. Instead, the six initial tokens on the arc from A to B in Figure 12.4a avoids a deadlock.

Figure 12.5 shows all the ways the system in Figure 12.4a can behave within a single cycle, which are exactly all the ways it can be scheduled. Starting from the state in the upper-right corner, there is only one choice: to run B. This consumes two of the tokens on the A→B arc and produces one token on the B→C arc and four on the B→D arc. At this point, there is a choice to run either B or D.

This graph is analogous to Figure 11.2, but obviously much simpler. The difference is not an accident: the communication patterns in SDF are far more regular because they do not depend on data. Communication within the Kahn process network in Figure 11.2 depends on the internal state of the processes; communication in Figure 12.5, and in any SDF system, depends only on the number of tokens in each buffer.

Although there are at most two choices at any point during the simulation of this system, there can be as many choices at there are actors depending on the topology of the system.

Any path through Figure 12.5 corresponds to a valid schedule. All paths fire C once, A twice, B three times, and D four times, matching the firing rates determined from the balance equation.

Two possible schedules are marked in Figure 12.5. One, BDBDB-CADDA, minimizes the maximum sizes of the buffers. Another, BB-BCDDDDAA, is easier to code but requires large buffers.

A common technique for generating code from SDF simply inlines the code for each actor. The most brute-force technique lays down one firing after another, producing, say, the implementation in Figure 12.6a for the schedule BBBCDDDDAA.

Such straightline code runs very fast, and is well-suited for implementation on a DSP. Code for each actor can be written by hand in

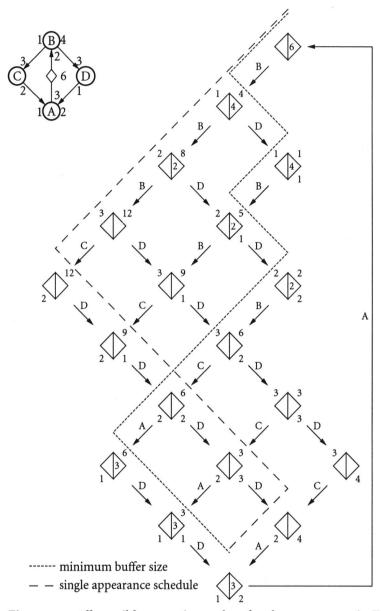

------- minimum buffer size
— — single appearance schedule

Figure 12.5: All possible execution orders for the SDF system in Figure 12.4a. Any path through this graph corresponds to a schedule.
Two schedules are shown.

```
B;
B;        for (i=0; i<3; i++)
B;           B;
C;        C;
D;        for (i=0; i<4; i++)
D;           D;
D;        for (i=0; i<2; i++)
D;           A;
A;
A;
```

(a)   (b)

Figure 12.6: Two implementations of the SDF system in Figure 12.4a. In (a), each invocation of a actor becomes a separate copy. A looped implementation (b) minimizes code duplication.

highly-tuned assembly language and later pieced together by the SDF scheduler. Function calls are generally not used because their overhead can easily dominate computation time for simple actors.

Using counted loops to avoid duplicating code is an obvious improvement. Figure 12.6b shows an example. When the size of the code for, say, D, is large, loops are much more compact and nearly as fast.

12.2   Looped Scheduling

In general, choosing schedules that trade off code size for buffer size is a challenging problem. Bhattacharyya, Murthy, and Lee's book [11] is the definitive reference on solving this problem for SDF; many of its results first appeared in papers [10, 12] and Bhattacharyya's thesis [13].

Looped schedules consist of sequences of actor firings in nested, counted loops. A shorthand way to write these schedules parenthesizes loops and adds a loop count. For example:

$$(3 \text{ B})\text{C}(4 \text{ D})(2 \text{ A}) = \text{BBBCDDDDAA}$$
$$(2 \text{ BD})\text{BCA}(2 \text{ D})\text{A} = \text{BDBDBCADDA}$$
$$(2 \text{ A}(2 \text{ B})\text{C})\text{D} = \text{ABBCABBCD}$$

The first and third examples here are single-appearance schedules, i.e., each actor appears exactly once in the concise representation. Neglecting loop overhead, an SAS produces the smallest code size of any implementation, but one does not always exist and it may require larger buffer sizes than other schedules. An SAS may also not be unique, leaving some room for reducing required buffer sizes.

The procedure for finding good looped schedules takes a divide-and-conquer approach. Single-appearance schedules always exist for acyclic graphs (the nodes appear in topological order), so the graph is first decomposed into strongly-connected components (sccs). Then, arcs that do not impose an execution order are removed. Such an arc has at least as much delay as the total number of tokens transferred across it during a complete cycle of the strongly-connected component, implying the arc could never cause a actor to wait for data because enough is always available. With luck, removing these arcs splits each strongly-connected component and the actor is repeated on these smaller components. At the end, each component is either a single actor or the smallest group of actors that does not have a single-appearance schedule.

For example, the system in Figure 12.4a has a single-appearance schedule because of the delay on A→B. The graph is a single strongly-connected component, but because the delay on A→B is at least as much as the number of tokens that will cross it during a complete iteration, the arc never forces A to fire so B can fire. Removing this arc leaves the graph acyclic and makes it possible to have a single-appearance schedule. Specifically, both (3 B)C(4 D)(2 A) and (3 B)(4 D)C(2 A) are valid single-appearance schedules.

In general, the scheduling algorithm consists of an acyclic scheduling procedure, a partitioning procedure, and a cyclic scheduling procedure. Each of these pieces may be selected independently from one of many variants to improve buffer size or other metrics.

Consider finding a single-appearance schedule for the SDF modem in Figure 12.7. First, most actors fire once per cycle; the exceptions are a and b (16) and c and n (2). There is one strongly connected component consisting of actors d, e, f, i, j, k, l, m, n, o, and p. Within this SCC, arcs e→d and f→i do not affect execution order because they each have as much delay as tokens that will pass through them per period. Both actors e and f fire once per iteration; e produces two tokens, equal to the delay on e→d, and f produces one, equal to the delay on f→i. Removing these two arcs breaks the strong connectivity of the scc, making it easy to schedule:

$$ijklm(2 \ n)dofpe.$$

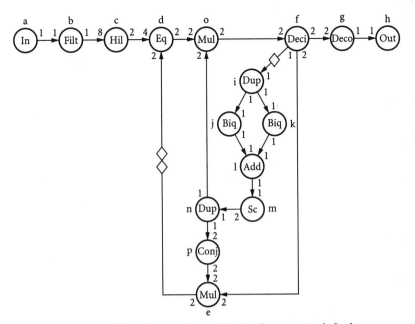

Figure 12.7: A modem in SDF. From Bhattacharyya et al. [12].

The schedule for the SCC is then inserted in the schedule for the whole graph, giving

$$(16\,a)(16\,b)(2\,c)\ ijklm(2\,n)dofpe\ gh.$$

A single-appearance schedule is not always unique. A collection of SCCS usually has many topological orders, each of which leads to a different schedule with different buffer size requirements.

12.3 Cyclo-Static Dataflow

A drawback of SDF is its insistence on strict actors, i.e., those that require all of their input before proceeding. For certain actors, such as an adder, this is a reasonable assumption. Something like a down-sampler, on the other hand, would be perfectly content to produce its output after seeing only a single sample.

Cyclo-static dataflow [14] (CSDF) is a variant of SDF that breaks the firing of an actor into phases to allow non-strict actors while remaining statically schedulable. In each phase, the actor behaves like an

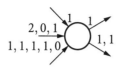

Figure 12.8: A CSDF actor with three input and two output ports. Labels indicate the communication pattern on each port: the number of tokens the actor consumes or produces each time it fires.

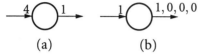

(a)   (b)

Figure 12.9: (a) An SDF down-sampler and (b) an equivalent CSDF actor.

SDF actor, producing or consuming a fixed number of tokens, but the communication pattern may be different in each phase. The balance equations are unchanged (the coefficients are the number produced or consumed in all phases).

The benefit of CSDF comes from its ability to fire actors in smaller steps. This usually reduces buffer sizes since groups of data can be handled piece by piece. CSDF systems are more suitable for parallel or pipelined implementations because their actors often produce data earlier than equivalent SDF actors.

Figure 12.8 shows a CSDF actor that has the same overall rate as the SDF actor in Figure 12.1. Each port has a label consisting of a sequence of integers. When the actor fires, each port produces or consumes one of these numbers of tokens and goes to the next number. For example, an input port labeled 2, 1, 3 would consume two tokens the first time its actor fires, one token the next time it fires, then three, then two, etc. It takes a actor as many firings as the least common multiple of lengths of the sequences on its ports to return to the same state. Figure 12.8 has sequences of length one, two, three, and five, so it repeats its behavior every thirty cycles.

Limiting the number of tokens produced and consumed per port per firing to one makes CSDF particularly well-suited for implementation in hardware. This limitation allows most buffers to be single-place and nicely matches the synchronous digital logic style. Such systems usually have a single clock that controls the whole system, and actors are designed to be able to fire completely in a single clock cycle. The scheduler becomes a simple state machine that signals to each actor whether to advance (fire) in each clock cycle. If an actor needs more than a clock cycle, the extra computing time can be hidden in phases that produce and consume no tokens.

Reconvergence, i.e., when two paths from the same node split and later join, can require larger buffers even when CSDF is limited to single tokens. Buffering is needed when the source produces data at the same time but the destination needs one input before the other.

12.4   Exercises

12–1. Draw an SDF actor for an $8\times$ upsampler.

12–2. What is the main advantage of SDF over Kahn Process Networks?

12–3. (a) Can an SDF system satisfy the balance equations and still not be able to run forever? (b) Can an SDF system violate the balance equations yet still run forever?

12–4. Why does SDF need initial tokens? (Channels in Kahn Process Networks always start empty.)

12–5. What does it mean about the structure of the system if its balance equations have a two-dimensional solution (i.e., one of of the form $\alpha q + \beta r$)? A three-dimensional solution?

12–6. Why is Figure 12.5 so much simpler than Figure 11.2? Figure 12.5 has no crossing lines, but is this possible for all SDF systems?

12–7. What is the main tradeoff among different looped schedules?

12–8. Why is a single-appearance schedule desirable for generating code?

12–9. True or false: If each strongly-connected component of a SDF system has a single-appearance schedule, then the system has one too. Why?

12–10. What is the main difference between SDF and CSDF? When would each be superior?

Part IV

Hybrid

These languages combine ideas from others to cover different parts of the design space. Esterel excels at discrete control by blending software-like control flow with the synchrony and concurrency of hardware. Polis employs extended finite-state machines communicating through single-place buffers to describe mixed hardware and software systems. Communication protocols are SDL's forte; it uses extended finite-state machines with single input queues. SystemC blends styles for software and hardware in C++ to provide a smooth way to refine software models into hardware. CoCentric System Studio combines dataflow with Esterel-like finite-state machine semantics to simulate and synthesize dataflow applications that also require control.

13

Esterel

The synchronous language Esterel [8, 7] can be thought of as a concurrent software language with hardware semantics. It is a textual imperative language whose constructs and syntax resemble a traditional sequential programming language such as C, but it has deterministic concurrency based on the notion of a hardware-like clock and instantaneous broadcast communication. It is well-suited for describing control-dominated hard real-time systems because it provides precise control over when events happen through high-level control constructs including preemption and exceptions.

An Esterel program is a collection of threads that communicate through signals. Execution is divided into cycles. At the beginning of each cycle, internal signals are reset to unknown and each thread resumes from where it paused at the end of the last cycle. Each thread runs until it hits a pause, occasionally waiting for other threads to establish the presence or absence of a signal. Threads can make a signal present by emitting it, but a signal is declared absent only when no thread is able to emit it in the current clock cycle. These semantics give Esterel deterministic concurrency.

Esterel can be implemented in hardware or software. Hardware is the easiest to understand: Esterel's clock becomes a global clock, each signal becomes a single wire, and each instruction becomes a small group of gates with an input that indicates control has reached the instruction and one or more outputs indicating how control leaves the instruction. A pause instruction becomes a register.

Compiling Esterel into software is less natural because the concurrency must be simulated. Fortunately, the concurrency is static and can be analyzed extensively at compile time. Current schemes translate an Esterel program into a single C function called once per cycle. The state of the program in the last cycle is stored in variables that this function uses to resume computation where it left off.

13.1 An Example

Figure 13.1 shows the traffic-light controller from Figure 2.11 implemented as three Esterel modules: a state machine, a timer, and a module that connects these two and runs them in parallel.

The FSM module takes input from the farm-road car sensor and two timeout signals from the timer to produce signals that indicate which of the four states the traffic lights are in. The body of the module consists of an infinite loop containing a sequence of emit and await statements. The await statement always waits for at least one cycle before checking that its signals are present. Here, this means at most one of the state signals is emitted in the same cycle.

Esterel deals with events better than values, so in this example a signal like HG only appears in one cycle. It is possible to use signals more like values (the SHORT signal, for example, behaves like this), but this is usually more clumsy to write.

The TIMER module takes a reset signal generated by the FSM and the SECOND signal to generate SHORT and LONG timeout signals. The module assumes the environment generates a SECOND signal once a second (this is not part of the language), and the module (perhaps more appropriately called a counter) uses this to determine delays.

The timer uses sustain statements to make the SHORT and LONG signals level-sensitive. When executed, sustain emits its signal every cycle forever. The FSM module expects SHORT and LONG to be level-sensitive because it includes a statement like await [CAR and LONG] that waits for CAR and LONG to be present in the same cycle. If LONG was present for a cycle, and CAR was present in the next, the await would still be waiting.

The timer uses a weak abort statement to restart whenever the FSM instructs it to. A weak abort terminates its body at the end of the cycle

```
module FSM:
input CAR, LONG, SHORT;
output RESTART;
output HG, HY, FG, FY;

loop
  emit HG ; emit RESTART; await [CAR and LONG];
  emit HY ; emit RESTART; await SHORT;
  emit FG ; emit RESTART; await [not CAR or LONG];
  emit FY ; emit RESTART; await SHORT;
end
end module

module TIMER:
input RESTART, SECOND;
output LONG, SHORT;

  loop
    weak abort
      await 3 SECOND;
      [
          sustain SHORT
      ||
          await 5 SECOND;
          sustain LONG
      ]
    when RESTART;
  end
end module

module TLC:
  input CAR, SECOND;
  output HG, HY, FG, FY;

signal RESTART, LONG, SHORT in
  run FSM
||
  run TIMER
end
end module
```

Figure 13.1: The traffic-light controller in Esterel

in which its signal is true. By contrast a strong abort (written simply abort) terminates its body at the beginning of the cycle, preventing its body from running.

The timer uses the weak abort to avoid a causality cycle. If the timer had instead used a strong abort, the timer would need to know the value of RESTART before it emitted SHORT or LONG. However, the FSM generates RESTART in response to these two signals. Had a strong abort been used here, the compiler would have reported this causality violation and rejected the program.

The abort statement in the timer illustrates the utility of statements that run forever. Such statements are generally enclosed in preemption constructs that can terminate them.

The timer illustrates how concurrency is often used in the small in Esterel. Once three seconds have passed, a pair of statements start: one that sustains SHORT and the other that waits five additional seconds before sustaining LONG. These two really are independent, so it is natural to specify them concurrently. Esterel encourages such micro concurrency; even in software such concurrency is very cheap.

The TLC module does nothing more than run the FSM and timer modules concurrently. The semantics of the run statement are closer to a function call than the concurrent instantiation of, say, Verilog. Although run is not recursive, it runs a module until the module terminates, at which point control continues from it. In this example, the two modules are run concurrently. If both were to terminate (they cannot in this example), the parallel statement would terminate and control would continue from the parallel.

Module ports are connected implicitly: unless otherwise specified, an input or output on a module is connected to a signal with the same name in the scope of the run statement. In this example, the two modules communicate between themselves through the signals RESTART, LONG, and SHORT, which are local to the TLC module.

13.2 Basic Signal-Handling Instructions

Esterel is built around the kernel of signal-handling instructions in Figure 13.2. The rest of the language consists of constructs that can be built from these and a set of data-handling instructions.

nothing	Do nothing
emit S	Make S present this instant.
present S then p else q end	Run p if S is present, q otherwise.
loop p end	Run p. Restart it when it terminates.
p ; q	Run p, wait for it to terminate, then run q.
pause	Stop here; resume in the next instant.
suspend p when S	Run p in this instant, but only in later instants when S is absent.
exit T	Throw exception T; leave that trap statement.
trap T in p end	Run p, terminating when p does or when exception T is thrown.
p || q	Run p and q concurrently; terminate when both do.

Figure 13.2: Esterel's kernel kernel statements. These handle pure signals and control flow.

The emit statement makes the given signal present in the one cycle in which the emit runs. This is the only statement that sets a signal (no statement makes a signal absent). It terminates instantly.

The present statement tests the presence or absence of a signal in the cycle in which it runs. Like all statements that read signals, this statement waits until emissions of the signal that could possibly execute in a cycle have run.

The loop statement is an infinite loop that must contain at least one pause. Traps can exit loops.

The semicolon runs statements in sequence: the second starts after the first terminates.

The pause statement is the only one that takes multiple cycles. Specifically, when control reaches a pause, its thread pauses and control resumes at the next statement in the next cycle.

The suspend statement prevents its body from executing in cycles when its signal is present. It immediately runs its body, but if the body does not terminate (i.e., waits at a pause), the suspend only allows its body to run in later cycles when the suspending signal is absent. To avoid causality problems, the body may not influence this signal.

The exit and trap statements generate and catch exceptions. The body of a trap statement runs in each cycle. If a corresponding exit within the body runs, the trap terminates once the body has finished (either at pauses or by terminating) for the instant. If the body contains two or more parallel threads and one runs an exit, the trap terminates every thread within its body. If two or more exit statements run, the outermost corresponding trap takes precedence.

Statements separated by the parallel operator || start in the same cycle. The group terminates the cycle when none of the threads are still running: one thread may have terminated while the other still runs; only when the latter terminates does the ensemble terminate.

13.3   Derived Statements

The pure kernel is powerful enough, but it can be clumsy. Esterel provides many derived statements that correspond to multiple kernel statements.

The halt statement stops without terminating:

```
halt                    loop pause end
```

The await statement is one of the most common. It pauses until the next cycle in which its signal is present:

```
await S                 trap T in
                          loop
                            pause;
                            present S then exit T end
                          end
                        end
```

A multiway variant of await waits for multiple signals at once. The earlier case wins if more than one is active.

```
await                   trap T in loop
  case SA do emit A        pause;
  case SB do emit B        present SA then emit A; exit T
end                        else present SB then emit B; exit T
                           else nothing
                        end end end
```

There is some asymmetry in the kernel statements. The trap statement controls the termination of its body and the end of a cycle; the suspend statement only prevents its body from running at the beginning of a statement. Together, these combine to build Esterel's symmetric set of preemption operators.

Strong abortion, where the presence of a signal terminates statement before it has a chance to run in an instant, can be built like this:

```
abort                     trap T in
   pause; pause              suspend pause; pause when S;
when S                       exit T
                          ||
                            loop
                              pause;
                              present S then exit T end
                            end
                          end
```

The loop with the present checks the condition and exits the trap when the signal is present. In that instant, the suspend statement prevents the body from running.

Weak abortion, on the other hand, terminates a statement after the condition appears. The weak abort statement does this:

```
weak abort                trap T in
   pause; pause              loop
when S                          pause;
                                present S then exit T end
                             end
                          ||
                             pause; pause; exit T
                          end
```

The loop-each statement restarts whenever a signal arrives. It is a loop wrapped around a strong abort.

```
loop                      loop
   pause; pause              abort
each S                          pause; pause; halt
                             when S
                          end
```

13.4 Reincarnation

Reincarnation is an odd aspect of the language. In certain cases a signal may appear to take two or more values in a single instant:

```
loop
  signal S in
    present S then emit O else nothing end ;
    pause ;
    emit S
  end signal
end loop
```

In the second instant, S is emitted, the `signal` statement terminates, and the loop resets with a fresh, absent copy of S. Signal O is not emitted. Detecting cases where multiple copies of the same signal may exist within the same instant is a subtle compilation problem, but can be determined statically.

13.5   Compilation Into Automata

First-generation compilers exhaustively simulate an Esterel program to generate a single large finite-state machine. Each state corresponds to a set of points where the program can pause between cycles and has a control-flow graph leading from it that decides the next state. Nodes in this graph emitted signals and test a signal's presence.

Very efficient code is easy to generate from a finite-state description. One technique translates each state into a C function and stores the machine state in a function pointer called once per cycle.

The automata approach can generate very large programs because it must explicitly list all combinations of behaviors. For example,

```
loop await A; emit C; await B; emit D end
||
  loop await E; emit F; await G; emit H end
||
  loop await I; emit J; await K; emit L end
```

has eight states since each thread has two states and all combinations are possible.

It can be practical to compile large programs if they are crafted carefully, i.e., if the number of states they contain is reduced by limiting the number of combinations that must be considered. Forcing threads to operate in a lockstep fashion greatly reduces the number of states. Reducing the number of input combinations that must be considered also helps. To aid this, Esterel provides a way to constrain a program's inputs. An input signal can be declared as implying another or a set may be declared mutually exclusive.

The compiler developed as part of the Polis project [5, 20] tries to share code between states. It combines the control-flow graphs from all the states in a single binary decision diagram (BDD) [16] which is guaranteed to share matching parts of the graph. The generated code is a large switch statement that branches on the state. The code for each state is a series of actions and conditions that often branches into the code for a different state. The result is hard to read because of all the gotos, but it is fast and compact. See Section 14.5.

Despite these tricks, most large programs simply cannot be compiled using automata techniques.

13.6   Compilation Into Boolean Circuits

In about 1992, Berry began experimenting with translating Esterel into logic gates [6]. Esterel and synchronous digital logic have essentially the same semantics, so the translation is fairly natural. Each halt point becomes a register. A one is loaded into the register when control should resume from that point in the next cycle. Conditionals become a pair of two-input AND gates. Each gate has one go input; the other is the condition or its complement. The outputs of these gates drive the go inputs of the instructions in the true and false branches.

The semantics of a circuit are concurrent and the notion of the currently-executing instruction is a loose one. For example,

```
present A then
  present B then
    nothing
  else
    emit C
  end
end
||
  emit B;
  present C then nothing end;
  emit A
```

appears to be a deadlock since the first thread produces C only after it knows A and B. The second thread only produces A once it knows the value of C.

Automata compilers use potentials to decide when signals can be declared absent. At any point, all emit statements reachable from that point without passing through a pause statement are considered to be runnable. Under these semantics, the above example would deadlock, since C depends on A and B, and A depends on C.

This example does not deadlock with circuit semantics. Emit B can run immediately, so the circuit concludes emit C cannot run since B is present. Since this is the only source of C, the circuit then concludes C is absent, allowing the second thread to emit A. Once the value of A is known the first thread finishes and the two threads terminate.

Capacity is the the main advantage of the circuit translation for software. Since each statement becomes just a few gates, the size of the circuit is linear in the size of the source program. This is much less than the potentially exponential size of automata code.

Once a netlist has been generated from Esterel source, software can easily be generated from it. Provided the netlist has no cycles, the gates it contains may be ordered (topologically sorted) such that each gate's inputs are computed before it must be. The generated program simply computes the function of each gate in this order.

Circuits generated from correct Esterel source, unfortunately, may have cycles. These are often false cycles in the sense that they may not be enabled in their entirety because the program cannot reach a state that would enable them. Consider

```
loop
  emit A; emit B; pause;
  emit B; emit C; pause;
  emit C; emit A; pause
end
||
  loop
    present [A and D] then emit E end; pause
  end
||
  loop
    present [B and E] then emit F end; pause
  end
||
  loop
    present [C and F] then emit D end; pause
  end
```

The last three threads appear to have a cyclic dependency (D $\rightarrow$ E $\rightarrow$ F $\rightarrow$ D). But the first thread ensures that at least one of A, B, or C is always absent, always breaking the dependency. But the circuit has a wire for each signal transmitted between threads, so it is cyclic. Moreover, no wire can just be removed since each is needed sometime.

The automata approach has no problems with such an example since it creates different code for each state, and does not require them to have the same order. Each state realizes that one of the three threads is inactive and simply does not generate code for it.

Berry's v4 compiler, which uses the circuit translation and requires the resulting circuit to be acyclic, rejects this example. Users, however, complained that these new semantics were too restrictive, so technology was developed to translate these cyclic circuits into acyclic ones.

Cyclic circuits produced from valid Esterel programs are combinational: they have a unique solution and do not oscillate, as would an inverter fed back on itself, or hold state, as would an SR latch built from a pair of cross-coupled NAND gates. Malik [56] studied this problem in hardware and devised a way to attach meaning to well-behaved cyclic circuits. The solution requires knowledge of the states the circuit can reach and proving the circuit is well-behaved in each. Shiple, Berry, and Touati [70] applied this technique to identify and compile well-behaved cyclic Esterel programs. The technique is exact, but can be computationally very expensive.

13.7   Compilation with Synthesized Context Switches

In 1999, I devised an new method for compiling Esterel into software [26, 27] that overcomes some of the inefficiencies of the automata and boolean logic methods. It translates Esterel source more or less-one-to-one into a concurrent control-flow graph that has a natural translation into software. This preserves the sequential nature of Esterel source where possible.

The compilation method relies on two main techniques. Nested multi-way branches (e.g., C's switch statements) resume the program where it paused in the last cycle. Concurrent behavior, in particular cases where threads must be interleaved because they communicate bidirectionally within a cycle, is simulated by inserting code that saves a thread's program counter in a variable and later restores it with a multi-way branch. In the end, the generated code looks more like what a human would write.

Since it requires source statements to be executed in the same order in each state, my compiler cannot compile programs that are statically cyclic but dynamically well-behaved.

13.8 Exercises

13–1. (a) What is a causality cycle? (b) Give an example in Esterel. (c) How would you break this cycle by changing the code? (d) Give an example of a false cycle, i.e., one that appears to be present but is not because the conditions that would trigger it can never occur.

13–2. What are the trade-offs among the three types of compilers? Why would you choose one over another?

13–3. Why did the language designers choose to make `await A` always wait a cycle before checking its signal?

14

Polis

The Polis hardware/software codesign system [5] is targeted toward the design, synthesis, and verification of control-dominated embedded hardware/software systems. A research project based primarily at the University of California, Berkeley, Polis has since formed the basis of a commercial product from Cadence.

The Polis system model, a network of codesign finite-state machines (CFSMs), was designed to describe control-dominated behavior that could be automatically translated to hardware and software. CFSMs are reactive finite-state machines with datapaths communicating through single-place buffers. CFSM transitions are atomic, but overall system behavior is asynchronous to model hardware and software running independently. Single-place buffers were chosen because they are simple to implement in both hardware (as a register) and software (as a memory location). To enable hardware synthesis and simplify software synthesis, the structure of the system (CFSMs and channels) is fixed while the system runs.

A CFSM network maps naturally to software as tasks running under an RTOS. Each CFSM plus its datapath becomes a task; each transition, which usually takes a short and fairly predictable amount of time, becomes a task invocation. The undefined speed and communication behavior of a CFSM matches the relatively uncontrolled behavior of a task running on an RTOS.

A hardware CFSM network behaves more predictably than its software equivalent, but maps just as easily. Each CFSM becomes com-

```
module COUNTDOWN:
input INITIAL : integer;
input SECOND;
output TIMEOUT;

every INITIAL do
  await ?INITIAL + 10 SECOND;
  sustain TIMEOUT;
end

end module
```

Figure 14.1: A countdown timer in Esterel. The arrival of the INITIAL signal resets the counter and sets the number of seconds (plus ten) to count. After the countdown, INITIAL is emitted constantly.

binational logic implementing its datapath and next-state function; the state variable and input ports all become latches. Since CFSMs are restricted to have non-zero delay, they are implemented as Moore-style FSMs whose outputs are latched and appear in the next cycle. This avoids the problems with zero-delay feedback that can arise on systems composed of Mealy machines, such as Esterel (Section 13.6).

14.1 CFSMs and the SHIFT Format

Polis represents a CFSM network using the textual software-hardware intermediate format (SHIFT, e.g., Figures 14.3 and 14.4). SHIFT uses KISS-like tables to define CFSMs and SPICE-like hierarchy to connect them. The format is intended to be machine-readable, so it is rather inscrutable. Fortunately, Polis provides mechanisms for generating SHIFT files from higher-level specifications.

Polis uses Esterel as a high-level language for CFSMs. Polis's Esterel-to-SHIFT translator uses an automata-based Esterel compiler (Section 13.5) to determine the states and transitions for the program, which becomes a single CFSM plus a datapath. A single CFSM is synchronous and a good match for Esterel semantics, but the asynchronous communication of a CFSM network makes it behave very differently than a collection of Esterel modules.

Polis translates the Esterel program in Figure 14.1 into a SHIFT file consisting of the finite-state machine in Figure 14.4 connected to the datapath in Figure 14.3. Figure 14.2 depicts this graphically.

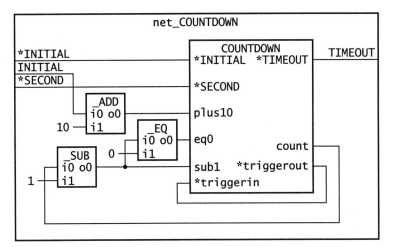

Figure 14.2: The structure of the SHIFT file for the countdown timer in Figures 14.3 and 14.4. The COUNTDOWN block is a finite state machine, and the _ADD, _SUB, and _EQ blocks comprise the datapath. The triggerout signal allows the state machine to fire repeatedly once the count has expired (to implement the Esterel sustain command).

14.2 Communication

The .net in Figure 14.3 defines how datapath functions and the CFSM from Figure 14.4 communicate. It lists ports for the countdown block, internal signals, and instantiates each of the sub-blocks, listing how their ports connect to internal signals.

CFSM networks are reactive: events cause computation. To support this model, CFSM networks communicate through single-driver, multiple-receiver channels that can convey events, values, or both. In Figure 14.3, the *SECOND input port conveys events (the leading * is a hint). The timer assumes there is an event on this channel once per second and counts them to measure time. The arrival of the *SECOND event initiates a CFSM transition, as would the arrival of any event.

The datapath portion of the countdown timer uses valued channels. For example, the CFSM generates the internal signal count (declared in the .internal list and marked as a value by the surrounding parentheses) to feed it to the subtractor (.subckt _SUB). The

```
.net net_COUNTDOWN
.inputs  *INITIAL(INITIAL), *SECOND
.outputs *TIMEOUT
.internal (count), (plus10), (eq0), (sub1), \
          (CONST_0), (CONST_1), (CONST_10), *trigger

.nb INITIAL            15 # Widths of valued variables
.nb count              15 # count has 16 bits
.nb plus10             15
.nb eq0        unsigned 1
.nb sub1               15
.nb CONST_0  unsigned  1
.nb CONST_1  unsigned  1
.nb CONST_10 unsigned  4

.const CONST_0   0     # Integer constants
.const CONST_1   1
.const CONST_10 10

# The Datapath: instances of built-in arithmetic operators

.subckt _ADD add i0=INITIAL i1=CONST_10 | o0=plus10
.subckt _EQ  eq  i0=sub1    i1=CONST_0  | o0=eq0
.subckt _SUB sub i0=count   i1=CONST_1  | o0=sub1

# Instance of the CFSM

.subckt COUNTDOWN countdown \
   *INITIAL=*INITIAL *SECOND=*SECOND \
   plus10=plus10 eq0=eq0 sub1=sub1 *triggerin=*trigger | \
   *TIMEOUT=*TIMEOUT count=count  *triggerout=*trigger
.end
```

Figure 14.3: The .net part of a SHIFT file for the program in Figure 14.1. This connects the CFSM in Figure 14.4 with its datapath as shown in Figure 14.2. A .net may also hierarchically compose other .nets.

```
.cfsm COUNTDOWN

.inputs *INITIAL, *SECOND, (plus10), (eq0), (sub1), *triggerin
.outputs *TIMEOUT, (count), *triggerout
.state (_st)

.nb plus10       15  # plus10 is 16 bits
.nb eq0 unsigned  1  # eq0 is a single bit
.nb sub1         15
.nb count        15
.mv _st 4 1 2 3 4     #_st has the four values 1, 2, 3, and 4

.r _st          1   # 1 is the initial state
.r *TIMEOUT     0   # The reset value of *o_TIMEOUT is 0
.r *triggerout  0

.trans                        # Transitions, one per line
#I S + = - t ps => T count    t ns
 - - - - - - 1    0 -          - 2 # Leave count unchanged}

 1 - - - - - 2    0 (plus10) - 3 # Copy plus10 to count
 0 - - - - - 2    0 -          - 2

 0 0 - - - - 3    0 -          - 3
 0 1 - 0 - - 3    0 (sub1)    - 3
 0 1 - 1 - - 3    1 (sub1)    1 4
 1 - - - - - 3    0 (plus10) - 3

 0 - - - - 1 4    1 -          1 4
 1 - - - - - 4    0 (plus10) - 3
.end
```

Figure 14.4: A CFSM in SHIFT. This is the state machine in the COUNTDOWN block in Figure 14.2. The count output can come from the plus10 or sub1 inputs, or remain unchanged.

line .nb count 15 sets the count signal to be sixteen bits wide (bits are numbered from zero to this number), the default for Esterel integers. The sub1 channel is an example of a channel with more than one receiver (here, the _EQ and COUNTDOWN blocks). A channel may have any number of receivers but must have exactly one driver. The .const CONST_1 line sets the driver of the CONST_1 to be a constant.

The *INITIAL(INITIAL) port conveys a valued event. Specifically, the *INITIAL event prompts the counter to reset itself and load itself with the INITIAL value plus ten.

Communication in a CFSM network is asynchronous and has undefined delays. Each input port stores the most recently-received event and value in a single-place buffer. This allows the transmitter to send data without having to wait for the receiver.

Buffering inputs instead of outputs matters when a event is sent to multiple receivers. An event buffer is a flag set by a transmitter sending the event and cleared by the receiver consuming it. Buffering inputs ensures each receiver has its own copy, avoiding a race where one receiver consumes an event before the other sees it.

Two rules improve the coherency of valued events. On a channel with valued events, the value must be written before the event and the event read before the value. Together, these ensure a receiver will never see an old value with a new event.

CFSMs may lose events and values because they run independently and communication channels have only single-place buffers. Nothing prevents a transmitter from sending a new event before the last one was consumed; a transmitter cannot even check for this without receiver cooperation.

14.3   Finite-State Machines

A CFSM begins to react when at least one event arrives on any of its inputs. Each reaction corresponds to a step of a traditional finite-state machine: the CFSM reads its inputs, locates a transition that matches the inputs and present state, clears the event flags on all its inputs, advances its state, and writes its outputs.

A CFSM transition is atomic, but its communication behavior is not. This is a compromise for modeling concurrently-running software. More synchronous communication would demand greater constraints on RTOS scheduling and the hardware software/interface. A reaction takes an unpredictable, non-zero amount of time, and reading and writing port data may occur in nearly any order. However, each input is read exactly once per transition.

Events on different input ports are not guaranteed to be coherent. For example, if another CFSM was generating both *INITIAL and *SECOND in the same transition, the countdown CFSM may see both events in the same transition, or one each in two transitions. The problem, fundamentally, is that a CFSM has no idea when all its inputs have arrived, so it is greedy about it: it consumes events as soon as it knows it has them, but since there is no control over how long it will take to respond to an event, other events may arrive before it has responded to the first one.

Because of the asynchronous communication mechanism, a single CFSM rarely sees simultaneous events. This differs from the Esterel philosophy, which assumes (and can guarantee) events often appear concurrently. For example, in Esterel is is natural to wait for a pair of signals to appear in the same instant, i.e.,

```
await [A and B]
```

In a CFSM, such a construct would probably fail. If another CFSM emitted these two signals in the same instant, it is possible that the receiving CFSM would not see them in the same instant. As a result, the receiver would consume one of the events in one reaction, the other in a second reaction, and not advance its state since it would not have seen both signals in the same instant. More disturbingly, the behavior of such a system is speed-dependent: if the CFSM reacts slower than the communication, then it might see both events in the same instant, but if the CFSM reacts more quickly, it will consume each event independently.

The solution is to allow the two signals to appear in any order, i.e.,

```
[await A || await B]
```

This construct has more states and will advance if A appears first, if B appears first, or both appear in the same instant.

A CFSM may be incompletely specified or nondeterministic. This is a natural consequence of a CFSM being fundamentally a set of transitions. An incompletely-specified CFSM is missing some transitions, i.e., not every possible combination of inputs and states has a matching transition. A nondeterministic CFSM has too many transitions, i.e., there is more than one matching transition for some combination of inputs and states.

When an incompletely-specified CFSM encounters an input condition with no matching transition, it does nothing. In particular, it does not consume the events on its inputs, and instead waits for more events to arrive to make some transition match. There is a possibility of deadlock here: if an input pattern has no matching transition, and no pattern with additional events has a matching transition, the CFSM will never react again.

For example, here is a CFSM that reacts non-monotonically.

```
module NONMONO:
input A, B; output C;
  await [A and not B or not A and B]; emit C;
end module
```

By default, the oc2shift translator (translates Esterel automata to SHIFT) generates the set of transitions

```
#A B ps => C ns
  1 1 1    0 1
  1 0 1    1 2
  0 1 1    1 2
  0 0 1    0 1
```

The first transition means the CFSM will consume both A and B if they ever occur together. By contrast, giving the -e flag to oc2shift makes it only generate transitions that do something, i.e.,

```
#A B ps => C ns
  1 0 1    1 2
  0 1 1    1 2
```

Here, if A and B ever occur simultaneously, the machine will deadlock: both events will remain in the input buffers and will never be consumed since the input/present state combination has no matching transition.

When a CFSM finds more than one matching transition, it is free to execute whichever it wishes. Naturally, simulating systems with non-deterministic CFSMs is problematic since the synthesized system may behave differently than the simulation by using a different rule to choose transitions. This is in addition to the nondeterminism from asynchronous communication.

CFSM transitions must be fairly simple. Only simple transitions can execute in a single cycle in hardware, and multi-cycle transitions would complicate the hardware synthesis process. Also, simple transitions make for short task times for an RTOS, requiring fewer time-consuming context switches and leading to more predictable behavior. Finally, simple transitions keep a CFSM represented with the table-based SHIFT file format to a reasonable size.

The CFSM in Figure 14.4 defines inputs and outputs, a state variable, the widths of its I/O ports, reset values for its state and outputs, and a table of transitions.

Inputs and outputs beginning with an asterisk, such as *INITIAL and *SECOND, convey events. An event is either present or absent, and it is the presence of an event that causes a CFSM to make a transition. Only CFSMs may manipulate events; combinational functions may not respond to or generate events.

An input or output such as (plus10) (enclosed in parenthesis) conveys a integer value whose width in bits is set by the .nb directive. Hence, the plus10 input is a sixteen-bit integer, the default translation for Esterel integers.

The .r directives set the state and the initial value of the CFSM's outputs. Outputs need initial values because a transition may leave an output's value unchanged. This is indicated, for example, by the - in the count column in the first transition.

The transitions are listed one per line after the .trans directive. Each has a pattern listing input values and present state followed by values for outputs and the next state.

Consider the second transition:

```
1 - - - - - 2     0 (plus10) - 3
```

The columns correspond, in order, to the inputs, present state, outputs, and next state. Thus, the 1 indicates the INITIAL input, an event, must be present for the transition to be taken. The next few –'s indicates this transition does not care about the presence or values of the other inputs (SECOND, plus10, etc.). The 2 indicates the transition starts in state 2. The 0 indicates the first output, the *TIMEOUT signal, is not to be emitted. The (plus10) indicates the value of plus10 will be copied to the count output when the transition is taken. The next – indicates the next output, *triggerout, is not to be emitted. Finally, 3 indicates to send the machine to state 3.

14.4   Hardware Synthesis

Translating a CFSM into hardware is straightforward. The structure of the synthesized circuit is essentially that of the CFSM (e.g., Figure 14.2). The implementation is synchronous; all CFSMs are driven off the same global clock. State bits, naturally, are latched. Outputs are also latched, ensuring a one-cycle delay between when an event arrives and when the machine can produce a reaction. Inputs are not buffered: in effect, the machine consumes all its events in each cycle. As such, the machine is forced to be completely specified.

Combinational logic for the datapath is built by instantiating arithmetic circuitry for each of the built-in functions (such as the adder and comparator in Figure 14.3) and connecting them as required by the .net. It is the designer's responsibility to ensure the combinational functions are simple and fast enough to complete in a single clock cycle.

Combinational logic for the finite-state machine is built in two pieces. A decoder generates a binary number that corresponds to the transition that matches the current input and present state. Each output is driven by a multiplexer that uses the output of the decoder to select what the value of the output will be.

14.5   Software Synthesis

Polis synthesizes a function for each CFSM destined for software, such as Figure 14.5. Software synthesis translates CFSM transitions into a binary-decision diagram (BDD), attempts to minimize the number of nodes in the BDD, and finally generates a control-flow graph from the structure of the BDD.

Divide-and-conquer is the basic idea. If you test an input that can be zero or one, some number of transitions match when that input is zero, others match when it is one. These groups of transitions, once split, can be split again on another variable, and so on until the values of the outputs are determined. Many of these groups of transitions are the same, so they can be combined. The result is a directed acyclic graph that either tests an input or writes an output at each node. Test nodes have two or more outgoing arcs.

The order in which the variables are tested affects the size of the code since it can change the number of transition groups that may be merged. To produce the C code in Figure 14.5, the synthesis procedure used the order INITIAL, st, triggerin, SECOND, count, EQ, TIMEOUT, triggerout, and st. Figure 14.6 shows the control-flow graph and a reordered list of transitions that illustrates the construction. The nodes labeled L3, L5, and L7 illustrate how nodes can be shared to reduce code size.

Overall, this techniques produces very fast code for small CFSMs. The generated code is nearly a lookup table: each variable is tested at most once, which is optimal for speed. For larger CFSMs, though, the generated code grows very quickly. Basically, the number of nodes is roughly the same as the number of transitions, a number that follows the number of states in the CFSM, which can be exponential in the size of the Esterel source. Nevertheless, this is not a significant impediment since Polis encourages very small CFSMs, since the difficulties in synthesizing large CFSMs in hardware are even more significant.

14.6   Communication Synthesis

Each CFSM may be in hardware or software and may communicate with other CFSMs in either hardware or software. Thus, there are four types of communication channel.

Hardware-to-hardware communication just uses wires. Since each CFSM runs synchronously and makes one transition per clock, events need only be held high for a single clock cycle and there is no need to latch events on the input. Instead, each CFSM output is latched.

Hardware-to-software communication ensures the RTOS reads an event by using a request/acknowledge protocol. A latch (Figure 14.7)

```
#define   _ADD(i0, i1)    ((i0)+(i1))
#define   _SUB(i0, i1)    ((i0)-(i1))
#define   _EQ(i0, i1)     ((i0)==(i1))
static unsigned_int8bit v__st = 0;

void _t_z_countdown(int proc, int inst)
{
    int _tmp;
    static int16bit v_INITIAL_tmp, v_count_tmp;
    static unsigned_int8bit _st_tmp;

    v_INITIAL_tmp = v_INITIAL; /* Make local copies */
    v_count_tmp = v_count;
    _st_tmp = v__st;
    startup(proc);
    if (INITIAL) goto L9;
    switch (_st_tmp) {
    case 2:  goto L4;
    case 3:  goto L8;
    default: goto L3;
    }
L3:  v__st = 1; goto L0;
L4:  if (SECOND) goto L6;
L5:  v__st = 2; goto L0;
L6:  v_count = _SUB(v_count_tmp,1);
     _tmp = _EQ(_SUB(v_count_tmp,1),0);
     if (!_tmp) goto L5;
L7:  emit_e_TIMEOUT();
     emit_e_trigger();
     v__st = 3;
     goto L0;
L8:  if (trigger) goto L7; else goto L0;
L9:  if (_st_tmp == 0) goto L3; else goto L10;
L10: v_count = _ADD(v_INITIAL_tmp,10); goto L5;
L0:  always_cleanup(proc); /* Consume all input events */
     return;
}
```

Figure 14.5: Generated C code for the countdown timer in Figures 14.1–14.4

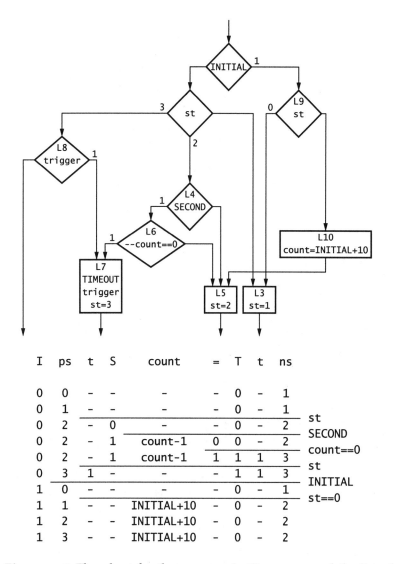

Figure 14.6: Flowchart for the program in Figure 14.5 and the list of transitions that generated it. Lines between transitions indicate places where decisions are made in the flowchart. For example, the transitions below the line labeled INITIAL are handled by the code under the true arc from the topmost node (labeled INITIAL).

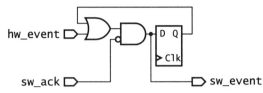

Figure 14.7: The circuit to communicate from hardware to software. A single cycle of hw_event sets the latch, holding sw_event high until the software acknowledges the event by raising sw_ack. sw_event goes to either an I/O value that the RTOS can poll or an interrupt.

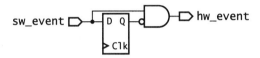

Figure 14.8: The circuit for communicating from software to hardware. Software sets the sw_event input high for many clock cycles before lowering it. This causes the hw_event signal to be high for one cycle, signaling a single event to a CFSM in hardware.

holds the event and transmits it to the processor. The output becomes an interrupt or an I/O location to be polled. In either case, some time elapses before the RTOS acknowledges the event by resetting the latch.

Software-to-hardware communication uses an edge detector (Figure 14.8). Generally, the software raises and lowers an I/O signal for some number of clock cycles. An edge detector watches for the rising edge of this signal and turns it into a high value for a single clock cycle: a single event in hardware.

Software-to-software communication writes to a CFSM input event buffer and tells the operating system the receiving CFSM is now ready to run. Polis synthesizes a function for each CFSM output that propagates the outputs to the appropriate places. For example, if the the emit_e_TIMEOUT signal is sent to a pair of CFSMs called MONITOR and OTHER, Polis would generate the function

```
void emit_e_TIMEOUT()
{
  deliver_e_TIMEOUT_to_z_MONITOR; /* Writes global variable */
  set_ready(_p_z_MONITOR); /* Inform RTOS */

  deliver_e_TIMEOUT_to_z_OTHER;
  set_ready(_p_z_OTHER);
}
```

14.7 Exercises

14–1. List two sources of nondeterminism in the Polis formalism.

14–2. Why did Polis choose single-place buffers instead of arbitrary-length ones?

14–3. Why buffer the inputs of a cfsm instead of its outputs?

15

SDL

SDL (Specification and Description Language) evolved from pencil-and-paper drawings used to describe the structure and behavior of large telecommunications systems. It formally began life in 1976 as a CCITT recommendation and has since been adopted and refined by the ITU into standard Z.100 [43], first released in 1988. The 1992 revision mainly added object-orientation; the 2000 revision improved object-oriented data modeling and added exception handling.

SDL describes distributed software systems as collections of blocks containing concurrently-running processes that communicate with signals sent through explicit channels. Each block represents a single processor, and each process it contains is an independent thread of control running on that processor. Processes in the same block communicate through zero-delay signal routes; blocks communicate through channels with delay.

Each process is an extended finite-state machine that reacts to arriving signals. A process has a single input queue, and in each state, a process consumes a signal (usually the oldest) from that queue, performs a series of actions in response to the signal (e.g., makes decisions, performs arithmetic, and sends signals), and advances to the next state. Each process runs when signals are waiting in its queue. Processes may be dynamically created and terminate while the system is running, but blocks are fixed.

SDL is best at specifying protocols: sequences of signals that arrive in an expected order. An SDL process handles such messages in se-

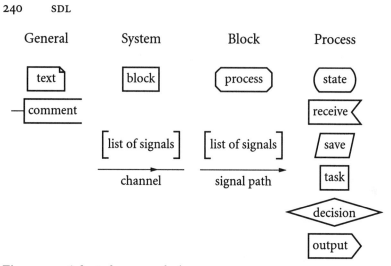

Figure 15.1: Selected SDL symbols

quence by receiving a signal, doing something in response to it, and waiting for the next. A process's input buffer prevents the loss of signals that arrive too quickly. Furthermore, each state may handle more than one signal, allowing for variations in a protocol (e.g., a signal may indicate which sequence of signals will arrive next).

SDL has both graphical and textual forms. The textual form is typical and resembles VHDL. The graphical form (Figure 15.1) uses block diagrams to represent communication between blocks and processes; processes are described with flowcharts. The graphical representation uses text annotations (⬜) for concepts, such as signal and type declarations, better suited to text.

The ultimate reference on SDL is the formal specification from the ITU [43], but Ellsberger et al. [29] and Olsen et al. [63] are better tutorials. The book edited by Turner [77] compares three formal description techniques including SDL.

15.1   An Example

Figures 15.2–15.5 show an implementation of the traffic light controller in SDL. It is more complicated than necessary to illustrate communication: it could be described as a single process.

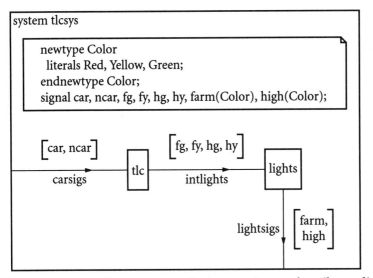

Figure 15.2: The traffic light controller system in SDL describes a distributed system. It contains two blocks and three channels. The text block defines enumerated and signal types. Each channel lists the signals it may convey in brackets.

15.2   Structure: Systems and Blocks

Figure 15.2 shows the top level system. Although overkill for such a small example (it could easily have been defined as a single block—SDL allows a block at the topmost level), it is modeled as a distributed system consisting of two blocks (processors), tlc and lights.

Information about whether a car is waiting at the farm road is conveyed by the car and ncar signals. SDL systems communicate through events, not values, so these two signals indicate when a car has just arrived at the sensor and when it has just left. These signals probably alternate, but the system does not require this.

The farm and high signals convey the most recent color of the farm road and highway traffic lights respectively. Since the two signals do not appear simultaneously, the lights process (Figure 15.6) is careful to make one light red before turning the other green.

To avoid bugs and simplify implementation, SDL communication pathways list (in brackets) the signals they convey. Each of the three channels in Figure 15.2 is also annotated with a name. The arrow in

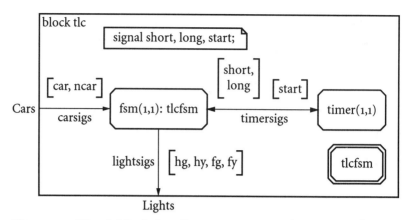

Figure 15.3: The tlc block describes processes running on a single CPU. It contains two processes and three signal routes. Each process lists the initial and maximum number of copies of it that can be running in (initial,maximum) form. The octagon in the lower right corner indicates the tlcfsm process is external.

the middle of each line indicates the channel has delay (the default at the system level, since it is assumed to be distributed).

A text box with a downturned corner is a textual annotation. The annotation in Figure 15.2 declares signals and a new enumerated type.

Figure 15.3 shows the tlc block used in the system. As all blocks do, this represents processes running concurrently on a single processor. This particular block contains the tlcfsm and timer processes.

Like the channels in Figure 15.2, the signal routes in Figure 15.3 are named and list the signals that can pass along them. The timersigs signal route is bidirectional: it has arrows at both ends and lists two groups of signals, one per direction.

The double octagon in the corner indicates the tlcfsm process is declared externally (unnecessary: done to illustrate the idea).

Each process instance lists the initial and maximum number of copies of it that can ever run. Here, both are marked (1,1), indicating exactly one will always run, but other markings are possible. A process that starts dormant but can be invoked while the system is running is marked (0,1); a process that handles a transaction might be marked (0), indicating there are none initially, but that any number may be created while the system is running.

15.3   Processes

Figure 15.4 shows the definition of the tclfsm process. Like all SDL processes, it is a finite-state machine drawn as a flowchart. The process starts at the state symbol ◯ without a label and immediately outputs (▭▷) the start and hg signals, which come from a different scope.

The tlcfsm process has six states: two wait for the short timeout (HY and FY, which handle the yellow lights); two pairs wait for a long timeout (HGn, HGc, FGn, and FGc, which handle green lights).

The state symbol ◯ plays three roles. Without a label, it denotes the entry point for the process (upper left corner). With an outgoing line, a state symbol defines a state (e.g., the topmost FGc). With an incoming arrow, it becomes a next state (e.g., FGc at the bottom). A state symbol (e.g., the topmost HGn on the left) may also be shared and have both incoming and outgoing arcs.

Receive symbols ▭◁ immediately follow states and indicate what to do when a process receives a signal. The receive on the left under the HGc state makes the process transition to the HGn state when it consumes the ncar signal in the HGc state; the receive on the right sends the start and hy signals before transitioning to the HY state.

15.4   The Save Construct

Normally, when a process receives a signal in a state that has no explicit rule for that signal, the signal is discarded. For example, the ncar signal will be discarded if it arrives in the HGn state.

Signals that arrive in a known order are easy to deal with since each process has a single input queue, but both the environment and the system itself, due to unpredictable communication delays, can produce signals in unpredictable orders. The save construct ▭ allows out-of-order signals to be handled in order. Attaching the save construct to a state keeps the saved signal in the queue and forces the process to look farther back in the queue for a signal to handle. This breaks the pure FIFO behavior of a process's input queue.

For example, state HY only handles the short timeout since the car and ncar signals are saved to be used by the FG states. The alternative would have been to introduce two HY states that keep track of whether car and ncar was seen most recently.

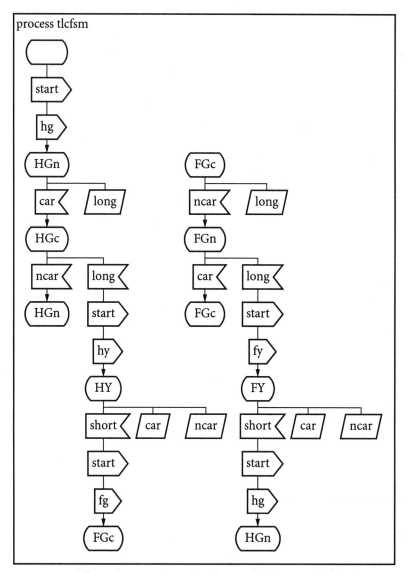

Figure 15.4: The tlcfsm process describes a single extended finite-state machine. The process begins at the empty state symbol in the upper left corner and emits the start and hg signals. A parallelogram (e.g., one labeled "long") indicates its signal is saved and not consumed in its state.

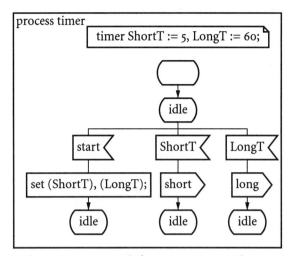

Figure 15.5: The timer process defines two timers, ShortT and LongT, and their default delays. The start signal resets them, and they emit signals when they time out.

15.5 Timers

Communication protocols, especially those that must operate over unreliable mediums, often use timeouts to handle cases when information has been lost. To assist with this, SDL provides timers. A timer can be thought of as an implicit process that can send a timeout signal to a process that requests it. Starting a timer begins a countdown that sends a signal after a specified period of time has elapsed.

The timer process in Figure 15.5 illustrates how timers are declared and used within SDL processes. A text box declares the two timers ShortT and LongT along with their timeout values (these are optional and can also be set while the process runs). When, say, the LongT timer times out, the timer sends a signal named LongT to the process, here causing the process to send the long signal to the tlcfsm process. The set action, here initiated by the start signal, restarts the timers.

This example split the timer and FSM into two processes to illustrate how processes communicate within blocks, but placing timers directly in the tclfsm process would lead to a more concise system.

Figure 15.6 shows the lights process, which turns four signals into two pairs of valued signals.

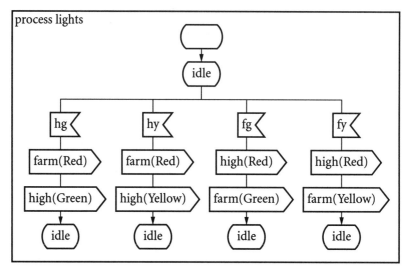

Figure 15.6: The lights process simply transforms one signal into two, so an external light controller receives all three colors for each light.

15.6   Exercises

15–1.   What is the main use of the save construct? How does it change the behavior of the input FIFO?

15–2.   What is the main difference between blocks and processes?

15–3.   What are the sources of nondeterminism in SDL?

15–4.   What advantages are there to labeling channels with the signals they convey?

16

SystemC

SystemC is a subset of C++ combined with a set of libraries aimed at the specification, simulation, and synthesis of digital hardware. It arose as a formal specification of what many designers had been doing for years to simulate very large systems, namely writing them in C or C++ in a structural style that would eventually lend itself to implementation in synchronous digital hardware.

One way to look at SystemC is a set of libraries that allows C++ to be written with Verilog or VHDL-like constructs: synchronous modules communicating through wires. SystemC has two main advantages over Verilog or VHDL. Simulation speed is the main advantage. A system simulated in SystemC can be simulated more efficiently than the equivalent in Verilog or VHDL because it does not have the overhead of event-driven semantics or exclusively four-valued datatypes. Timing details are lost, but most simulation time is spent checking timing-independent system-level behavior anyway.

The other advantage of SystemC is the refinement path it provides. Previous practice was to build a large simulation in C, verify it, then manually translate that specification into Verilog or VHDL, verify it again, and synthesize it into hardware, verify it again, etc. SystemC avoids the discontinuity by removing the hardware languages from the methodology. The same syntax and development tools are used throughout the refinement process.

A SystemC model is built from hierarchical modules that may contain processes and other modules. Modules and ultimately processes communicate through signals passed through ports on a module. Signals generally behave like single-driver wires; resolved signals are provided for modeling tristate busses. In addition to the types provided by C++ (integers, structures, etc.), SystemC supplies types for describing hardware, including fixed-point and four-value types.

SystemC provides three types of processes. A synchronous-thread process models synchronous logic: a clock starts one running, and it runs until it hits a wait() call. An asynchronous-function process models purely combinational logic by recomputing its outputs (run a C++ method to completion) whenever any of the signals in its sensitivity list (generally its inputs) change. Asynchronous-thread processes have both behaviors: they react to changes on their inputs, but hold control state between invocations. No hardware behaves quite like this; instead these processes are useful in testbenches.

Synchronous processes behave as if they have implicit latches on their outputs. Specifically, it is only after all synchronous processes triggered by a particular clock edge have run that any of their output signals are updated. This guarantees behavior will not depend on the (undefined) order in which synchronous processes run, but also prohibits two synchronous processes from communicating with each other in the same cycle; their communication would have to take multiple cycles. Such semantics sidestep the causality problems caused by connecting "zero-delay" modules (e.g., such as those in Esterel), and makes it easier to solve system timing at the module level. Unfortunately, these semantics can also make it impossible to decompose some behaviors into multiple processes. Behavior demanding same-cycle communication must be implemented in a single process.

16.1   An Example

Figure 16.1 shows a complex number multiplier modeled as an asynchronous function process. It declares four integer input ports, two integer output ports, a method for computing the function of the process, and a constructor that creates the asynchronous-function process and its sensitivity list.

While the specification is running, any change on signals a, b, c, or d causes the do_mult method to be invoked. This method reads

```
#include "systemc.h"

struct complex_mult : sc_module {
  sc_in<int> a, b;        // Integer-valued input and output ports
  sc_in<int> c, d;
  sc_out<int> x, y;

  void do_mult() {        // Perform the complex multiply
    x = a * c - b * d;
    y = a * d + b * c;
  }

  // Register do_mult as an asynchronous function process
  // and declare its sensitivity list

  complex_mult(const char *name) : sc_module(name) {
    sc_async_fprocess(handle, "mult", complex_mult, do_mult);
    sensitive << a << b << c << d;
    end_module();
  }
};
```

Figure 16.1: A complex number multiplier modeled as an asynchronous process in SystemC. When any signals on its sensitivity list change (i.e., a, b, c, or d) the do_mult function is invoked and computes the output values x and y.

the four input ports and writes the two output ports. The intention is that such a method only observes the signals on the sensitivity list and does not attempt to hold state between invocations. These restrictions help to ensure the module can be realized in hardware.

16.2   Modules

A SystemC module resembles those in Verilog or VHDL. It may have ports, internal signals and variables, processes, and instances of other modules. Syntactically, a SystemC module is a class derived from the sc_module class with a constructor that instantiates processes and other modules. Ports and signals are the module's data members.

Figure 16.2 shows a module (called hier_module) that contains three instances of two other modules within it. Like any module, a hierarchical module has ports through which it communicates, but it also may have internal signals used to connect instances of other modules it contains.

Within the class, each instance is a data member that points to the actual instance. Instances are created in the module's constructor and each is given a name.

Ports on each instance are connected to internal signals or ports in one of two ways. For modules with few ports, positional connection is the most convenient, otherwise, each port on an instance can be instructed to connect itself to a particular signal or port.

16.3   Processes

Figure 16.3 illustrates a module that creates one of each type of process supported by SystemC.

The three functions that create processes take four or five arguments. The first four arguments are a variable created automatically and set to a handle for the process, the name of the process (unique to the module), the module class, and the name of the method implementing the body of the process. For synchronous processes, the fifth argument is clock edge to which they will respond.

The body of a thread process (either synchronous or asynchronous) generally contains wait() and wait_until() statements that wait for the next clock or a condition.

```
struct moda : sc_module {
  sc_in<int>    a_iport;
  sc_out<float> a_fport;
  /* ... */
};

struct modb : sc_module {
  sc_in<char>  b_cport;
  sc_in<float> b_fport;
  sc_out<int>  b_iport;
  /* ... */
};

struct hier_module : sc_module {
  sc_in<char>   h_cport;    // I/O ports
  sc_in<int>    h_iport;
  sc_out<float> h_fport;

  sc_signal<float> h_fsig; // Signals connect internal modules
  sc_signal<int>   h_isig;

  module1 *inst1;           // Instances of modules within this
  module1 *inst2;
  module2 *inst3;

  hier_module(const char *name) : sc_module(name) {
    inst1 = new moda("a_inst1");    // Create and name
    inst2 = new moda("a_inst2");    // instances of modules
    inst3 = new modb("b_inst3");

    inst1->a_iport(h_isig);         // Connect by port name
    inst1->a_fport(h_fsig);

    (*inst2)(h_iport, h_fport);     // Connect by position

    (*inst3)(h_cport, h_fsig, h_isig);

    end_module();                   // Compulsory
  }
};
```

Figure 16.2: Instantiating modules within a module

```
struct example_module : sc_module {
  sc_in<int> porta;          // Ports
  sc_out<sc_bit> portb;
  sc_inout<float> portc;
  sc_in_clk clock;           // Synchronous process's clock

  sc_signal<int> sig1;       // Internal signals
  sc_signal<double> sig2;

  int var1;                  // Internal variables: do not
  char var2;                 // share among processes

  void proc1(); // Async. function: should not call wait();
  void proc2(); // Async. thread:   may call wait();
  void proc3(); // Sync. thread:    may call wait();

  example_module(const char *name) : sc_module(name) {

    // Asynchronous function process
    sc_async_fprocess(handle1, "p1", example_module, proc1);
    sensitive << porta << portb;

    // Asynchronous thread process
    sc_async_tprocess(handle2, "p2", example_module, proc2);
    sensitive(portb); sensitive(portc);  // alternate notation

    // Synchronous thread process: no sensitivity list needed
    sc_sync_tprocess(handle3, "p3", example_module, proc3,
                  clock.pos());  // Positive-edge sensitive

    end_module();
  }
};
```

Figure 16.3: An example illustrating how processes are instantiated within a module

```
#include "systemc.h"

struct complex_mult : sc_module {
  sc_in<int>  a, b;  // Integer-valued input and output ports
  sc_in<int>  c, d;
  sc_out<int> x, y;
  sc_in_clk   clock;

  void do_mult() {         // Perform the complex multiply
    for (;;) {
      x = a * c - b * d;
      wait();              // wait a cycle
      y = a * d + b * c;
      wait();              // wait a cycle
    }
  }

  // Register as a positive-edge sensitive synchronous process

  complex_mult(const char *name) : sc_module(name) {
    sc_sync_tprocess(handle, "mult", complex_mult, do_mult,
                     clock.pos());
    end_module();
  }
};
```

Figure 16.4: A complex multiplier modeled as a synchronous process in SystemC. By declaring it a `sc_sync_tprocess` the process will pause at the `wait()` statements within the body until the next positive edge of the `clock` input clock.

`sc_bit`	0 or 1
`sc_logic`	0, 1, X, or Z
`sc_bv<N>`	fixed-length vector of 0's and 1's
`sc_lv<N>`	fixed-length vector of 0's, 1's, X's, and Z's
`sc_int<N>`	fixed-precision integer (<64 bits)
`sc_uint<N>`	fixed-precision unsigned integer (<64 bits)
`sc_bigint<N>`	fixed-precision integer (≥64 bits)
`sc_biguint<N>`	fixed-precision unsigned integer (≥64 bits)
`sc_fixed<N,I>`	fixed-point number (N bits, I integer)
`sc_ufixed<N,I>`	fixed-point unsigned number (N bits, I integer)

Figure 16.5: SystemC's built-in types for hardware modeling

The `wait_until` construct, like `wait`, suspends the execution of its process, but it waits for the next time a condition is true. These conditions involve ports and signals, e.g.,

```
wait_until(a.delayed == 1 && b.delayed == 0);
```

Processes must communicate among themselves using local signals instead of using class variables (a single process may use its own internal variables) because the order in which processes are executed is undefined. If two processes, say, modify a variable in an order-dependent way, simulation results may change unexpectedly.

16.4   Types

In addition to C++ built-in types (integer, floating-point, character, etc.), SystemC provides additional types well-suited to hardware modeling (Figure 16.5).

Bit vectors are designed to model digital signals that cannot be undefined (e.g., uninitialized) or undriven (e.g., a bus). Generally, they simulate much faster than their four-valued counterparts.

Two types of fixed-precision integers are provided, one suited for less than sixty-four bits, the other for larger numbers. The two are provided for efficiency reasons.

Fixed-precision integers and fixed-point numbers are well-suited for hardware implementation since they simply correspond to some number of wires in a bus (unlike software, where there are generally a few natural word sizes and anything larger or smaller is difficult to implement). Many signal-processing algorithms are implemented in hardware with fixed-point numbers since they provide a convenient way to balance numerical precision with hardware resources.

16.5   Ports, Signals, and Clocks

Scalar ports (all the examples so far) are declared with `sc_in<T>`, `sc_out<T>`, or `sc_inout<T>`, where T is either one of the built-in C++ types, one of the SystemC types, or a user-defined type that includes the == operator.

SystemC also provides array port and signals for modeling multi-bit busses, such as address or data values. The syntax is similar:

```
sc_in<int>     a[10]; // Ten input ports: a[0], ..., a[9]
sc_signal<char> c[16]; // Sixteen signals: c[0], ..., c[15]
```

For more detailed hardware modeling, SystemC provides ports of bit vectors (each value may be zero or one) or logic vectors (may be zero, one, X, or Z). For example:

```
sc_in_bv<16> ain;     // 16-bit vector port of ones and zeros
sc_out_bv<2> bout;
sc_inout_bv<4> bio;
sc_signal_bv<8> bvsig;

sc_in_lv<17> lvin;   // Vector of 17 0, 1, X, or Z's
sc_out_lv<2> lvout;
sc_inout_lv<3> lvio;
sc_signal_lv<8> lvsig;
```

Resolved logic vectors are used for modeling hardware busses that use tristate drivers to select which one is currently driving the bus. These are declared as follows:

```
sc_in_rv<16> rin;  // Resolved vector of 16 0, 1, X, or Z's
sc_out_rv<8> rout;
sc_inout_rv<4> rio;
```

In general, plain types simulate faster than bit vectors, which simulate faster than logic vectors, which simulate faster than resolved logic vectors.

1. Update the clock or clocks that change at this simulation time.

2. Update the outputs of all newly-triggered synchronous processes (those that are sensitive to a clock that has changed) and remember them for step 5.

3. Execute all asynchronous processes whose inputs have changed in some order. (They may generate clocks for synchronous processes.)

4. Repeat steps 2 and 3 until no signal has changed its value.

5. Execute all the synchronous processes triggered in step 2 in some order. They do not update their outputs at this point.

6. Advance simulation time to the next clock and repeat.

Figure 16.6: SystemC's scheduling algorithm

Clocks are special signals used by SystemC to trigger synchronous processes and keep track of time. A new clock signal is declared by listing its name, period, duty cycle, delay to first edge, and sense of the first edge (`true` means positive, `false` negative). For example, a clock with period of 20 time units, a 50% duty cycle, whose first edge is positive and starts at 3 time units after the start of simulation is declared with:

```
sc_clock my_clock("myClock", 20, 0.5, 3, true);
```

A clock input port is declared with `sc_in_clk`; an output with `sc_out_clk`. Since each clock has exactly one driver, there is no need for a inout clock port.

16.6   Scheduling SystemC

SystemC uses cycle-based simulation semantics to avoid the overhead of an event queue. This is natural for modeling system-level behavior and generally much faster because timing information is ignored.

Figure 16.6 shows the scheduling algorithm, which operates in two phases. The first phase alternates between updating the outputs of synchronous processes and executing asynchronous processes that respond to those outputs. The second phase evaluates all the synchronous processes that saw a clock change in the first phase.

Overall, this simulates zero-delay synchronous digital logic. Delaying changing the outputs of synchronous processes until the next clock cycle mimics the behavior of edge-sensitive flip-flops. Overall, this ensures the nondeterministic order of synchronous process execution in step 5 does not matter: if the processes cannot see each other's outputs until the next cycle, the order in which they are executed is irrelevant.

Implementing the wait() statement is the main trick in SystemC. The presence of this operator turns a function into a coroutine, something not directly supported by C++. Instead, a threads package saves and restores processor state at wait() statements. Liao, Tjiang and Gupta describe the implementation [54] (Scenic was a code name for SystemC).

16.7 Exercises

16–1. What are the three types of processes in SystemC and why were they included?

16–2. How would you write each of the three types of SystemC processes in Verilog?

16–3. Why is there not a delay operator in SystemC? How does its absence simplify the scheduling algorithm?

16–4. Synchronous processes run in a nondeterministic order, yet SystemC is generally deterministic. What did the language designers do to ensure this possible?

17

CoCentric System Studio

CoCentric System Studio (once called El Greco—see Buck and Vaid-yanathan [17]) is a graphical design environment for systems built with a combination of dataflow diagrams and hierarchically-nested finite-state machines. It combines Esterel-like synchronous semantics for FSMs with Kahn's or cyclo-static dataflow semantics for dataflow.

System Studio distinguishes itself by allowing arbitrary nesting of FSMs and dataflow while retaining a formal, deterministic semantics. The technique for heterogeneous nesting follows the Ptolemy system [19]. The hierarchical finite-state machine semantics and notation resemble Andre's SyncCharts [2] which is based on Harel's Statecharts [35]. Dataflow uses Kahn's semantics [44] plus ideas from the cyclo-static dataflow model [14].

17.1 An Example

Figures 17.1 and 17.2 show a primitive cruise control unit described in System Studio. It illustrates the four types of hierarchical models: dataflow, OR, AND, and gated. The top of Figure 17.1 is a (trivial) dataflow model consisting of an instance of a single model, namely the OR model cc_top drawn beneath it. In this case, this model does little more than define inputs and outputs.

The cc_top model is an OR model: a group of mutually-exclusive states with predicated transitions between them. If the states of an OR model are all atomic, it behaves like a traditional FSM. But states

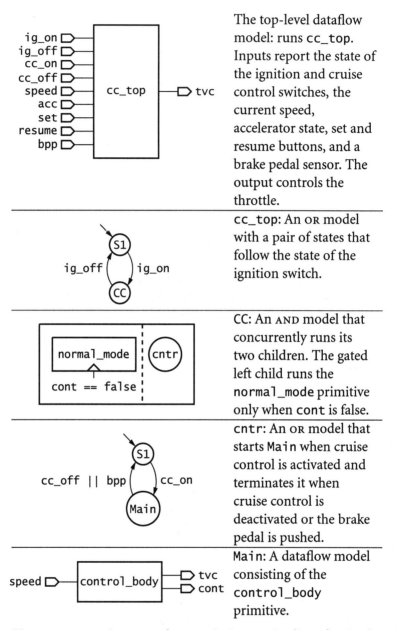

The top-level dataflow model: runs `cc_top`. Inputs report the state of the ignition and cruise control switches, the current speed, accelerator state, set and resume buttons, and a brake pedal sensor. The output controls the throttle.

`cc_top`: An OR model with a pair of states that follow the state of the ignition switch.

CC: An AND model that concurrently runs its two children. The gated left child runs the `normal_mode` primitive only when `cont` is false.

`cntr`: An OR model that starts `Main` when cruise control is activated and terminates it when cruise control is deactivated or the brake pedal is pushed.

`Main`: A dataflow model consisting of the `control_body` primitive.

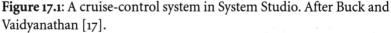

Figure 17.1: A cruise-control system in System Studio. After Buck and Vaidyanathan [17].

```
prim_model normal_mode {
  port in float acc;        // accelerator pedal position
  port out float tvc;       // throttle value control
  main_action {
    read(acc);
    tvc = acc;
    write(tvc);
  }
}

prim_model control_body {
  // I/O ports
  port in float speed;      // current speed
  port out float tvc;
  port out bool cont;

  // Parameter
  param float k1, k2;       // controls reaction times

  // Local Variables
  float speed_ref;          // reference speed: set in first cycle
  bool snap_ref = true;
  float tvc_int = 0.0;      // last throttle value setting

  main_action {
    read(speed);
    if (snap_ref) {         // Remember reference speed
      speed_ref = speed;    // in the first cycle
      snap_ref = false;
    }
    float delta = speed - speed_ref;
    tvc = tvc_int = k1 * delta + k2 * tvc_int;
    write(tvc);
    cont = true;
    write(cont);
  }
}
```

Figure 17.2: Code for the two primitive models in Figure 17.1. The normal_mode model simply copies the state of the accelerator pedal to the throttle valve, effectively bypassing the cruise control unit when it is inactive. The control_body model does the real work: it samples the speed when it starts then forever sets the throttle to try to achieve this speed.

may also contain other models, such as the CC state, which contains an AND model. When control enters a state containing a model, it resets the model before starting to run it. Such a model runs until a transition sends control to another state.

In a OR model, a transition may cause control to leave a state either before or after its contained model runs for the cycle. These two choices are called strong and weak termination, and provide control over the "fencepost" situation that arises when an exceptional condition occurs. In this example, the strong termination arc between CC and S1 makes shutting off the ignition (ig_off) take precedence over simultaneously turning on the cruise control (cc_on).

Although a model can weakly terminate itself, a model attempting to strongly terminate itself is a contradiction. If the model had run and made the strong termination predicate true, the model would not have been allowed to run.

OR models may have local variables that can be tested and updated as a side-effect of a transition or state. This allows, for example, counters to implemented without their states having to be listed explicitly.

The CC model is an AND model, which has a collection of child models that run and communicate synchronously using Esterel-like unbuffered semantics. Like Esterel, instantaneous cycles are possible and considered an error. They are flagged when the system is compiled. This particular AND model runs a gated model (on the left) and a OR model on the right. Implicitly, the signals generated by each child is broadcast to all other children in the same cycle, i.e., AND models use synchronous, unbuffered communication.

The gated model (left child of CC) runs the normal_mode primitive model (shown in Figure 17.2) whenever the cont signal (generated by the control_body primitive) is false. When the signal is true, the normal_mode model holds its state and does not run. Such gated models are good for situations where you want to switch between multiple operating modes (cruise control enabled and disabled in this example) without losing their state. Such gated models behave a little like a pause button.

In general, gated models may have two children, one that runs when a predicate is true, the other running when the predicate is false.

The cntr OR model illustrates how more complicated predicates can be placed on transitions. Here, the Main state terminates if either the cruise control turns off or the break pedal is pressed.

Finally, the Main model is a trivial dataflow model much like the topmost one. It consists of an instance of a single model, in this case the control_body primitive model shown in Figure 17.2.

17.2   Models

The interface of any model in System Studio consists of communication ports and parameters. Ports provide the only source of inter-model communication (i.e., System Studio does not support shared global variables). Each port has a type and a direction. Dataflow models have input and output ports only; control models may also have inout ports that allows the bidirectional flow of information. The semantics of inout ports parallels those in Esterel. In a dataflow context, a port represents a connection to a FIFO buffer; in control models, a port represents a binding (i.e., without buffering) to external signals.

Parameters come in many flavors. Type parameters, such as the one in Figure 17.3 allow for a limited form of polymorphism that allow the same design to be used with integers and fixed-point numbers.

Non-type parameters correspond to read-only data and come in three flavors. Structural parameters are defined when the system is compiled and do not change while the system is running. These are natural for things like bit widths that must be fixed before a system is fabricated, especially for hardware. Read-on-reset parameters are useful when a model is terminated and restarted repeatedly while the system is running. The parameter in Figure 17.3 that sets the down-sampling factor is typical. This is an infrequently-changing value on which few communication resources should be spent. Finally, dynamic parameters may change while a model is running. These are typically used for values that also change slowly and asynchronously, such as filter coefficients.

17.3   Dataflow models

System Studio's dataflow models have semantics like Kahn process networks but do not permit dynamic process creation. Each block in

```
prim_model DownSample {
  type_param T = float;
  port in T InData;
  port out T OutData;
  param read_on_reset unsigned Factor;
  main_action {
    for ( unsigned i = 1 ; i <= Factor ; i++ )
      read(InData);
    OutData = InData;
    write(OutData);
  }
}
```

Figure 17.3: A downsampler in System Studio's primitive language

a dataflow model may be any of the four hierarchical models or a primitive model. In a dataflow model, each instance of a AND, OR, or gated model is treated as unit-rate: each time it is invoked, each input consumes one token and each output produces one.

System Studio allows dynamic dataflow behavior (i.e., arbitrary data-dependent communication patterns), but attempts to schedule as much as possible statically for efficiency reasons. In particular, an attempt is made to statically identify the communication patterns of primitive models.

17.4   Primitive models

Primitive dataflow models are written in a C++-like syntax that permits full Kahn-like semantics (deterministic but essentially arbitrary communication patterns) but in many cases the communication can be analyzed and broken into phases and scheduled as cyclo-static dataflow. This is for simulation efficiency and whether the system can successfully perform this analysis does not change the behavior.

Consider the downsampler in Figure 17.3. It has input and output ports, a type parameter that allows it to be used polymorphically, a parameter that sets the downsampling factor (read when the model is reset), and a body that reads a number of input values before copying the last one to the output port.

The type parameter T allows the model to be reused with different data types (e.g., it could be used equally well to downsample floating-

point or fixed-point streams). This is a C++ template-style polymorphism that permits a type to be passed into the model when it is instantiated. Once the type is bound, the operators (e.g, the assignment
of InData to OutData) are replaced with type-specific versions.

The Factor parameter is marked read_on_reset, meaning it is
re-read every time the model is reset. This mechanism is convenient
for controlled dataflow models where the result of some computation (say, a communication-rate negotiation) is used to affect how a
dataflow computation behaves.

The model in Figure 17.3 can be statically scheduled because the
communication pattern only depends on the read-only parameter
Factor and not on data. Specifically, there is a phase group of Factor
reads of the input port followed by a single phase that writes to the
output port.

17.5   Exercises

17–1.   List the four types of models used in System Studio.

17–2.   Design a very small system in System Studio that has a causality error (hint: make an OR model state try to terminate itself).
How could you slightly change the behavior of your system to
avoid this problem?

17–3.   What good are System Studio's type parameters?

17–4.   Does a model within an OR state hold its state when control
leaves it?

Internet Resources

This chapter is intended to be a list of Internet resources related to the languages in this book (compilers, simulators, etc.). Sadly, the dynamic nature of the Internet will probably render many of these out-of-date by the time you read this.

Freely Available Software

Free software development tools abound; free hardware design tools are harder to find. Furthermore, many free hardware development tools are research prototypes that only address, for example, restricted subsets of a language.

   Simulator, schematic, and printed-circuit board layout tools, along with assemblers for embedded processors are more common for Windows. Chip-design software is more common on Unix platforms.

Hardware

`http://collector.hscs.wmin.ac.uk/`
> Open Collector at the University of Westminster is a very extensive database of freely available hardware design software for Unix and Windows.

`http://www.iserv.net/~alexx/lib/general.htm`
> Alex's Electronic Resource Library has general electrical information as well as a list of DOS/Windows software.

`http://www.geda.seul.org/`
> gEDA is a schematic capture system for Unix.

http://xcircuit.sourceforge.net

    Tim Edwards' XCircuit is a schematic capture system for Unix geared toward analog circuits.

ftp://ic.eecs.berkeley.edu/pub/Spice3

    The SPICE analog circuit simulator: this site has sources, others have precompiled binaries and ports to other architectures.

http://icarus.com/eda/verilog

    Stephen Williams' Icarus Verilog is among the best free Verilog simulators. It handles both structural and behavioral subsets and can generate netlists for Xilinx field-programmable gate arrays.

http://daggit.pagecreator.com/ver/ver.html

    The Ver Structural Verilog compiler supports a smaller subset than Icarus.

http://www-asim.lip6.fr/alliance/

    The Alliance project (Pierre et Marie Curie University in Paris) concentrates on IC design and supplies, among other things, a structural VHDL simulator.

Software

http://www.gnu.org

    The GNU project supplies excellent C and C++ compilers (gcc and g++), a portable assembler for many processors (part of the binutils package), and many, many other software development tools.

http://www.cs.princeton.edu/software/lcc

    lcc is the portable C compiler described in Fraser, Hanson, and Hansen [30].

http://java.sun.com

    Sun Microsystems' Java site provides Java development kits including compilers and interpreters.

`http://www.blackdown.org`
> The Blackdown project has ported Java development kits to the Linux operating system.

`http://www.redhat.com/services/ecos`
> RedHat's eCos (formerly Cygnus's) is a portable, open-source real-time operating system that has been ported to ARM, PowerPC, i386, SuperH, and SPARClite processors.

`ftp://ftp.cs.berkeley.edu/ucb/nachos`
> The NACHOS timesharing operating system was developed for instructional use at the Univeristy of California, Berkeley. Silberschatz and Galvin [71] includes an appendix on NACHOS written by Thomas Anderson.

Dataflow

`http://ptolemy.berkeley.edu`
> Ptolemy [18] is a highly-developed graphical entry and simulation system particularly good at Synchronous Dataflow modeling.

Hybrid

`http://www.esterel.org/`
> Esterel Web supplies compilers and more information about the language.

`http://www-cad.eecs.berkeley.edu/~polis`
> The Polis system for hardware/software codesign [5] (University of California, Berkeley and elsewhere) has estimation, simulation, and synthesis capabilities.

`http://www.sdl-forum.org/`
> The SDL forum has more information about SDL.

`http://www.systemc.org`
> The Open SystemC Initiative supplies SystemC simulation libraries and documentation.

Commercial Software

There is a big electronic design automation industry that supplies rather expensive CAD tools (they often donate to schools), and a far larger software development tools industry.

Hardware

`http://www.synopsys.com/`
> Large EDA company best known for its logic synthesis products, but also supplies compiled VHDL and Verilog simulators.

`http://www.cadence.com/`
> Large EDA company best known for physical design tools (IC layout). Supplies Verilog-XL, descended from the original Verilog simulator, and NC Verilog, a compiled simulator.

`http://gluon.fi.uib.no/~bruce/eda.htm`
> Bruce McKibben's Electronic Design Automation links: Mostly commercial EDA companies.

`http://www.ovi.org`
> Open Verilog International is a professional group devoted to promoting the Verilog language. Their site sells standards and provides access to working groups.

Software

`http://www.windriver.com/`
> Wind River System sells VxWorks, a successful RTOS.

Dataflow

`http://www.bdti.com`
> Berkeley Design Technologies reviews DSP processors and has some free information.

Hybrid

`http://www.telelogic.com`
Telelogic's Tau system is a SDL entry, simulation, and synthesis environment.

`http://www.itu.int`
The ITU's website sells documentation on standards such as SDL.

`http://www.synopsys.com`
Synopsys's CoCentric System Studio is a graphical entry and simulation environment for mixed control/dataflow specifications.

Acronym Glossary

ALU Arithmetic Logic Unit. A circuit element able to perform arithmetic operations such as addition. A central part of most processors.

ANSI American National Standards Institute. Source of many language standards including C and C++.

ARM Advanced RISC Machines. Company that sells an embedded microprocessor design that can be incorporated into integrated circuits.

ASCII American Standard Code for Information Interchange. A common computer character set. Now, usually augmented with characters such as ä and ç used in other Latin languages.

BCPL A simple typeless language designed in 1966 by Martin Richards and implemented at MIT. A main predecessor of C.

BDD Binary Decision Diagram. A compact way of representing binary functions. Used in logic synthesis, verification, and in Polis for software synthesis. A directed acyclic graph representation where each node is labeled with a variable and two outgoing arcs, one for each possible value of the variable.

BSD Berkeley Standard Distribution. A variant of the Unix operating system.

CCITT Comité Consultaif International Télégraphique et Téléphonique. Group responsible for many telecommunication standards including the SDL language and many voice compression standards. Now known as the ITU-T.

CD Compact Disc. An optical disc used for storage of music and data. Able to store 650 MB or 74 minutes of 16-bit stereo sound sampled at 44.1 kHz.

CFSM Codesign Finite-State Machine. A finite-state machine with a datapath: the main type of entity used in the Polis codesign system.

CISC Complex Instruction Set Computer. A processor with a complicated machine language better suited for human programming, such as the Intel x86 or Motorola 68000 architectures. Compare with RISC.

CMOS Complementary Metal-Oxide Semiconductor. The most common fabrication technology for digital integrated circuits. Allows both p- and n-channel devices on the same die.

CSDF Cyclo-static Dataflow. An SDF variant that fires actors in phases, each of which may or may not consume or produce tokens on certain arcs.

CPU   Central Processing Unit. A stored-program computer that executes
instructions.

DAC   Digital-to-Analog Converter. A circuit that produces a varying voltage or
current according to a digital input.

DSP   Digital Signal Processor. A CPU optimized for digital signal processing
algorithms.

DVD   An optical disc technology whose discs are the same size as Compact Discs.
Able to store up to 8 Gb. Once meant Digital Video Disc and Digital
Versatile Disc, but now stands alone.

EDA   Electronic Design Automation. The use of computer programs in the design
of electronic systems.

EDF   Earliest Deadline First. A scheduling technique that assigns higher priorities
to tasks that must complete earlier.

FFT   Fast Fourier Transform. A signal processing algorithm that transforms a
time-domain signal into its frequency-domain representation.

FIFO   First-In First-Out. A storage element that can be read or written. Data
values are read out in the same order that they are written.

FIR   Finite Impulse Response. A type of filter that responds for a finite amount
of time to a short input stimulus. Implemented by convolving the input
with the impulse response.

FSM   Finite-State Machine. A machine that can be in one of a finite number of
states. The machine moves between these states according to input values.

GNU   Gnu's Not Unix. A free software project that has produced many software
development tools such as compilers and editors.

HTTP   Hypertext Transfer Protocol. The networking protocol used for
communicating with world-wide web servers.

IC   Integrated Circuit. Silicon substrate on which transistors and wires have
been etched. A "Computer Chip."

IEEE   Institute of Electrical and Electronics Engineers. A professional
organization that, among other things, produces standards for languages
and communication protocols.

IIR   Infinite Impulse Response. An efficient digitial filter structure containing
feedback that responds indefinitely to a brief input and has non-linear
phase. Compare with FIR.

ILP   Instruction-Level Parallelism. The ability to initiate many instructions
concurrently. Assumes they do not depend on each other.

IP   Internet Protocol. A low-level networking protocol that understands packets
and addresses but not the concept of a connection or reliable delivery.

ITU   International Telecommunication Union. Standards body responsible for
SDL among others.

JVM Java Virtual Machine. The fictitious processor on which Java Bytecodes are executed.

KISS Keep Internal States Simple. A program developed by Giovanni De Micheli that encoded symbolic states. Also the input format to the program that describes finite-state machines.

MAC Multiply and Accumulate. A common operation in digital signal processors. This multiplies two numbers and adds the result to a register.

MIPS Million Instructions Per Second. A common measure of processor speed. Also the name of a common RISC processor.

MOSFET Metal-Oxide Semiconductor Field Effect Transistor. The transistor most commonly used in digital integrated circuits. Used as a voltage-controlled switch.

NACHOS Not Another Completely Heuristic Operating System. A timesharing OS designed for teaching developed at the Univeristy of California, Berkeley.

NAND Not-And. A common logic function where the output is false only if all the inputs are true.

NOP No-operation. An assembly-language instruction that does nothing. Often used as a placeholder or to flush the instruction pipeline.

NOR Not-Or. A common logic function where the output is true only if both the inputs are false.

OS Operating System. The program generally responsible for controlling peripherals, scheduling processes, and other system-level tasks.

OVI Open Verilog International. A standards group devoted to maintaining the Verilog standard and related standards.

PLA Programmable Logic Array. A regular two-level structure for implementing logic functions. Computes a sum (OR) of products (AND).

PLI Programming Language Interface. A set of functions and definitions that can be used to connect a Verilog simulator to a C program.

ROM Read-Only Memory. A common type of memory that can only be read. Its contents are set when it is manufactured. Commonly used to store programs and unchanging data in embedded systems.

RISC Reduced Instruction-Set Computer. A type of assembly language where the instructions are simple (typically, load, store, and perform arithmetic on registers) and regular. Examples include the SPARC architecture and the processors from ARM. Contrast with CISC.

RTL Register Transfer Level. A style of Verilog or VHDL in which data is moved between registers at fixed clock cycles.

RTOS Real-Time Operating System. A multitasking operating system that schedules processes to meet deadlines.

SAS Single Appearance Schedule. A looped schedule for SDF where each process appears once. Produces implementations with the minimum code size.

SCC   Strongly-Connected Component. In a directed graph, this is a subgraph in which there is a directed path between any two nodes. May be a cycle or may be more connected.

SDF   Synchronous Dataflow. A language for describing multirate digital signal processing algorithms.

SDL   Specification and Description Language. A graphical language that describes finite-state machines communicating through buffers. Well-suited to telecom systems.

SHIFT   Software-Hardware Intermediate Format. Represents CFSM's in Polis.

SPARC   Scalable Processor Architecture. A RISC processor architecture developed by Sun Microsystems.

SPICE   Simulation Program with Integrated Circuit Emphasis. An analog circuit simulator. Also the netlist file format it and many other tools read.

STA   Static Timing Analysis. An automated technique for checking the timing behavior of a circuit. Treats the circuit as a graph and find the longest path. Much faster than simulation, but prone to give false negatives.

STL   Standard Template Library. The collection of polymorphic data structures (arrays, trees) that is part of the C++ standard library.

UDP   User-Defined Primitive. A logic gate or sequential element defined with a truth table in the Verilog language.

VHDL   VHSIC Hardware Description Language.

VHSIC   Very High Speed Integrated Circuit.

VLIW   Very Long Instruction Word. A processor where each assembly instruction contains many fields controlling independent, concurrently-running instruction units. Difficult to program by hand, but can be very powerful.

XNOR   Exclusive Not-Or. Output is true if an even number of inputs is true.

XOR   Exclusive Or. Output is false if an even number of inputs is true.

Bibliography

Each entry ends with a list of the pages on which it is cited.

[1] A. Aho, R. Sethi, and J. Ullman. *Compilers, principles, techniques, and tools.* Addison-Wesley series in Computer Science. Addison-Wesley, Reading, Massachusetts, 1988. ⟨3, 134, 135⟩

[2] Charles André. Representation and analysis of reactive behaviors: A synchronous approach. In *Proceedings of Computational Engineering in Systems Applications (CESA)*, pages 19–29, Lille, France, July 1996. http://www-sop.inria.fr/meije/esterel/syncCharts/ ⟨259⟩

[3] Peter J. Ashenden. *The Designer's Guide to VHDL.* Morgan Kaufmann, San Francisco, California, 1996. ⟨55⟩

[4] Peter J. Ashenden. *The Student's Guide to VHDL.* Morgan Kaufmann, San Francisco, California, 1998. ⟨55⟩

[5] Felice Balarin, Paolo Giusto, Attila Jurecska, Claudio Passerone, Ellen Sentovich, Bassam Tabbara, Massimiliano Chiodo, Harry Hsieh, Luciano Lavagno, Alberto Sangiovanni-Vincentelli, and Kei Suzuki. *Hardware-Software Co-Design of Embedded Systems: The POLIS Approach.* Kluwer, Boston, Massachusetts, 1997. ⟨219, 223, 269⟩

[6] Gérard Berry. Esterel on hardware. *Philosophical Transactions of the Royal Society of London. Series A*, 339:87–104, 1992. ⟨219⟩

[7] Gérard Berry. The constructive semantics of pure Esterel. Book in preparation, 1996. ftp://cma.cma.fr/esterel/constructiveness.ps.gz ⟨211⟩

[8] Gérard Berry and Georges Gonthier. The Esterel synchronous programming language: Design, semantics, implementation. *Science of Computer Programming*, 19(2):87–152, November 1992. ftp://cma.cma.fr/esterel/BerryGonthierSCP.ps.Z ⟨211⟩

[9] Jayaram Bhasker. *A VHDL Synthesis Primer.* Star Galaxy Publishing, Allentown, Pennsylvania, second edition, 1998. ⟨55⟩

[10] Shuvra S. Bhattacharyya and Edwards Ashford Lee. Looped schedules for dataflow descriptions of multirate signal processing algorithms. *Journal of Formal Methods in System Design*, 5(3):183–205, December 1994. http://ptolemy.eecs.berkeley.edu/publications/papers/94/looped_schedules ⟨203⟩

[11] Shuvra S. Bhattacharyya, Praveen K. Murthy, and Edward Ashford Lee. *Software Synthesis from Dataflow Graphs*. Kluwer, Boston, Massachusetts, 1996. ⟨203⟩

[12] Shuvra S. Bhattacharyya, Praveen K. Murthy, and Edward Ashford Lee. Synthesis of embedded software from synchronous dataflow specifications. *Journal of VLSI Signal Processing Systems*, 21(2):151–166, June 1999. http://ptolemy.eecs.berkeley.edu/publications/papers/99/synthesis/ ⟨203, 204⟩

[13] Shuvra Shikhar Bhattacharyya. *Compiling Dataflow Programs for Digital Signal Processing*. PhD thesis, University of California, Berkeley, July 1994. Available as UCB/ERL M94/52. ⟨203⟩

[14] Greet Bilsen, Marc Engels, Rudy Lauwereins, and J. A. Peperstraete. Cycle-static dataflow. *IEEE Transactions on Signal Processing*, 44(2):397–408, February 1996. ftp://ftp.esat.kuleuven.ac.be/pub/acca/GRAPE/REPORTS/g95-04.ps.Z ⟨205, 259⟩

[15] Loic P. Briand and Daniel M. Roy. *Meeting Deadlines in Hard Real-Time Systems: The Rate Monotonic Approach*. IEEE Computer Society Press, 1999. ⟨183⟩

[16] Randal E. Bryant. Graph-based algorithms for boolean function manipulation. *IEEE Transactions on Computers*, C-35(8):677–691, August 1986. ⟨219⟩

[17] Joseph Buck and Radha Vaidyanathan. Heterogeneous modeling and simulation of embedded systems in El Greco. In *Proceedings of the Eighth International Workshop on Hardware/Software Codesign (CODES)*, San Diego, California, May 2000. ⟨259, 259⟩

[18] Joseph Tobin Buck, Soonhoi Ha, Edward Ashford Lee, and David G. Messerschmitt. Ptolemy: A mixed-paradigm simulation/prototyping platform in C++. In *Proceedings of the C++ At Work Conference*, Santa Clara, CA, November 1991. http://ptolemy.berkeley.edu/ ⟨199, 269⟩

[19] Joseph Tobin Buck, Soonhoi Ha, Edward Ashford Lee, and David G. Messerschmitt. Ptolemy: A framework for simulating and prototyping heterogeneous systems. *International Journal of Computer Simulation*, 4:155–182, April 1994. http://ptolemy.eecs.berkeley.edu/papers/JEurSim/index.html ⟨259⟩

[20] Massimiliano Chiodo, Paolo Giusto, Attila Jurecska, Luciano Lavagno, Harry Hsieh, Kei Suzuki, Alberto Sangiovanni-Vincentelli, and Ellen Sentovich. Synthesis of software programs for embedded control applications. In *Proceedings of the 32nd Design Automation Conference*, pages 587–592, San Francisco, California, June 1995. ftp://ic.eecs.berkeley.edu/pub/HWSW/dac95.ps.gz ⟨219⟩

[21] Ben Cohen. *VHDL Coding Styles and Methodologies*. Kluwer, Boston, Massachusetts, second edition, 1999. ⟨55⟩

[22] Giovanni De Micheli. *Synthesis and Optimization of Digital Circuits.* McGraw-Hill, New York, 1994. ⟨26, 31, 52⟩

[23] Giovanni De Micheli, Robert K. Brayton, and Alberto Sangiovanni-Vincentelli. Optimal state assignment for finite state machines. *IEEE Transactions on Computer-Aided Design,* CAD-4(3):269–285, jul 1985. ⟨27⟩

[24] Srinivas Devadas, Abhijit Ghosh, and Kurt Keutzer. *Logic Synthesis.* McGraw-Hill, New York, 1994. ⟨26, 52⟩

[25] Allen M. Dewey. *Analysis and Design of Digital Systems with VHDL.* Brooks/Cole Publishing (Formerly PWS), Pacific Grove, California, 1997. http://www.pws.com ⟨55⟩

[26] Stephen A. Edwards. Compiling Esterel into sequential code. In *Proceedings of the 7th International Workshop on Hardware/Software Codesign (CODES),* pages 147–151, Rome, Italy, May 1999. Association for Computing Machinery. http://www.sigda.acm.org/Archives/ProceedingsArchives/ ⟨221⟩

[27] Stephen A. Edwards. Compiling Esterel into sequential code. In *Proceedings of the 37th Design Automation Conference,* Los Angeles, California, June 2000. ⟨221⟩

[28] Margaret A. Ellis and Bjarne Stroustrup. *The Annotated C++ Reference Manual.* Addison-Wesley, Reading, Massachusetts, 1990. ⟨140⟩

[29] Jan Ellsberger, Dieter Hogrefe, and Amardeo Sarma. *SDL: Formal Object-Oriented Language for Communicating Systems.* Prentice Hall, Upper Saddle River, New Jersey, second edition, 1997. ⟨240⟩

[30] Christopher W. Fraser, David R. Hanson, and David Hansen. *A Retargetable C Compiler: Design and Implementation.* Addison-Wesley, Reading, Massachusetts, 1995. http://ftp.cs.princeton.edu/software/lcc/ ⟨135, 268⟩

[31] Erich Gamma, Richard Helm, Ralph Johnson, and John Vlissides. *Design Patterns: Elements of Reusable Object-Oriented Software.* Addison-Wesley, Reading, Massachusetts, 1995. ⟨140⟩

[32] Carl A. Gunter. *Semantics of Programming Languages.* MIT Press, Cambridge, Massachusetts, 1992. ⟨3⟩

[33] Gary D. Hachtel and Fabio Somenzi. *Logic Synthesis and Verification Algorithms.* Kluwer, Boston, Massachusetts, 1996. ⟨26, 52⟩

[34] Samuel P. Harbison and Guy L. Steele. *C: A Reference Manual.* Prentice Hall, Upper Saddle River, New Jersey, fourth edition, 1994. ⟨113⟩

[35] D. Harel. Statecharts: A visual formalism for complex systems. *Science of Computer Programming,* 8(3):231–274, June 1987. ⟨259⟩

[36] Randolph E. Harr and Alec G. Stanculescu, editors. *Applications of VHDL to Circuit Design.* Kluwer, Boston, Massachusetts, 1991. ⟨55⟩

[37] John L. Hennessy and David A. Patterson. *Computer Architecture: A Quantitative Approach.* Morgan Kaufmann, San Francisco, California, second edition, 1996. ⟨84, 91⟩

[38] J. Hopcroft and J. Ullman. *Introduction to Automata Theory, Languages, and Computation.* Addison-Wesley, Reading, Massachusetts, 1979. ⟨26⟩

[39] Paul Horowitz and Winfield Hill. *The Art of Electronics.* Cambridge University Press, second edition, 1989. ⟨19⟩

[40] R. D. M. Hunter and T. T. Johnson. *Introduction to VHDL.* Kluwer, Boston, Massachusetts, 1996. Formerly published by Chapman & Hall. ⟨55⟩

[41] IEEE Computer Society, 345 East 47th Street, New York, New York. *IEEE Standard VHDL Language Reference Manual (1076-1993),* 1994. http://www.ieee.org/ ⟨55⟩

[42] IEEE Computer Society, 345 East 47th Street, New York, New York. *IEEE Standard Hardware Description Language Based on the Verilog Hardware Description Language (1364-1995),* 1996. http://www.ieee.org/ ⟨32, 53⟩

[43] International Telecommunication Union, Place des Nations, CH-1221, Geneva 20, Switzerland. *ITU-T Recommendation Z.100: Specification and Description Language,* 1999. http://www.itu.int ⟨239, 240⟩

[44] Gilles Kahn. The semantics of a simple language for parallel programming. In *Information Processing 74: Proceedings of IFIP Congress 74,* pages 471–475, Stockholm, Sweden, August 1974. North-Holland. ⟨189, 189, 259⟩

[45] Timothy Kam, Tiziano Villa, Robert K. Brayton, and Alberto Sangiovanni-Vincentelli. *Synthesis of Finite State Machines: Functional Optimization.* Kluwer, Boston, Massachusetts, 1996. ⟨26⟩

[46] R. M. Karp and R. E. Miller. Properties of a model for parallel computations: Determinacy, termination, and queueing. *SIAM Journal on Applied Mathematics,* 14(6):1390–1411, November 1966. ⟨197⟩

[47] Randy H. Katz. *Contemporary Logic Design.* Benjamin/Cummings, Redwood City, California, 1994. ⟨19, 26⟩

[48] Brian W. Kernighan and Dennis M. Ritchie. *The C Programming Langage.* Prentice Hall, Upper Saddle River, New Jersey, 1978. ⟨133⟩

[49] Brian W. Kernighan and Dennis M. Ritchie. *The C Programming Langage.* Prentice Hall, Upper Saddle River, New Jersey, second edition, 1988. ⟨113⟩

[50] Donald E. Knuth. *The Art of Computer Programming,* volume one. Addison-Wesley, Reading, Massachusetts, third edition, 1997. ⟨119⟩

[51] Edward Ashford Lee and David G. Messerschmitt. Static scheduling of synchronous data flow programs for digital signal processing. *IEEE Transactions on Computers,* C-36(1):24–35, January 1987. ⟨199, 200⟩

[52] Edward Ashford Lee and David G. Messerschmitt. Synchronous data flow. *Proceedings of the IEEE,* 75(9):1235–1245, September 1987. ⟨197, 199⟩

[53] Samuel J. Leffler, Marshall Kirk McKusick, Michael J. Karels, and John S. Quarterman. *The Design and Implementation of the 4.3BSD UNIX Operating System.* Addison-Wesley, Reading, Massachusetts, 1989. ⟨179⟩

[54] Stan Liao, Steve Tjiang, and Rajesh Gupta. An efficient implementation of reactivity for modeling hardware in the scenic design environment. In *Proceedings of the 34th Design Automation Conference*, Anaheim, California, June 1997. ⟨257⟩

[55] C. L. Liu and James W. Layland. Scheduling algorithms for multiprogramming in a hard real-time environment. *Journal of the Association for Computing Machinery*, 20(1):46–61, January 1973. ⟨181⟩

[56] Sharad Malik. Analysis of cyclic combinational circuits. *IEEE Transactions on Computer-Aided Design*, 13(7):950–956, July 1994. ⟨221⟩

[57] M. Morris Mano. *Digital Design*. Prentice Hall, Upper Saddle River, New Jersey, second edition, 1991. ⟨19, 26⟩

[58] Marshall Kirk McKusick, Keith Bostic, Michael J. Karels, and John S. Quarterman. *The Design and Implementation of the 4.4BSD Operating System*. Addison-Wesley, Reading, Massachusetts, second edition, 1996. ⟨179⟩

[59] Carver Mead and Lynn Conway. *Introduction to VLSI Systems*. Addison-Wesley, Reading, Massachusetts, 1980. ⟨25⟩

[60] Swapnajit Mittra. *Principles of Verilog PLI*. Kluwer, Boston, Massachusetts, 1999. ⟨32⟩

[61] Steven S. Muchnick. *Advanced Compiler Design and Implementation*. Morgan Kaufmann, San Francisco, California, 1997. ⟨135⟩

[62] Lawrence W. Nagel. SPICE2: A computer program to simulate semiconductor circuits. Technical Report ERL M520, University of California, Berkeley, May 1975. ⟨21⟩

[63] Anders Olsen, Ove Færgemand, Birger Møller-Pedersen, Rick Reed, and J. R. W. Smith. *Systems Engineering Using SDL-92*. North-Holland, 1994. ⟨240⟩

[64] John K. Ousterhout. *Tcl and the Tk Toolkit*. Addison-Wesley, Reading, Massachusetts, 1994. ⟨84⟩

[65] Thomas M. Parks. *Bounded Scheduling of Process Networks*. PhD thesis, University of California, Berkeley, 1995. Available as UCB/ERL M95/105. http://ptolemy.eecs.berkeley.edu/ ⟨194, 194, 194⟩

[66] David A. Patterson and John L. Hennessy. *Computer Organization and Design: The Hardware/Software Interface*. Morgan Kaufmann, San Francisco, California, second edition, 1997. ⟨91⟩

[67] Douglas L. Perry. *VHDL*. McGraw-Hill, New York, third edition, 1998. ⟨55⟩

[68] P. J. Plauger. *The Standard C Library*. Prentice Hall, Upper Saddle River, New Jersey, 1992. ⟨113⟩

[69] Dennis M. Ritchie. The development of the C language. In *History of Programming Languages II*, Cambridge, Massachusetts, April 1993. http://www.cs.bell-labs.com/who/dmr/index.html ⟨131⟩

[70] Thomas R. Shiple, Gérard Berry, and Hervé Touati. Constructive analysis of cyclic circuits. In *Proceedings of the European Design and Test Conference*, pages 328–333, Paris, France, March 1996. ftp://ic.eecs.berkeley.edu/pub/Memos_Conference/edtc96.SBT.ps.Z ⟨221⟩

[71] Abraham Silberschatz and Peter B. Galvin. *Operating System Concepts*. Addison-Wesley, Reading, Massachusetts, fifth edition, 1998. ⟨179, 269⟩

[72] Bjarne Stroustrup. *The Design and Evolution of C++*. Addison-Wesley, Reading, Massachusetts, 1994. ⟨140, 159⟩

[73] Bjarne Stroustrup. *The C++ Programming Language*. Addison-Wesley, Reading, Massachusetts, third edition, 1997. ⟨140⟩

[74] Stuant Sutherland. *The Verilog PLI Handbook*. Kluwer, Boston, Massachusetts, 1999. ⟨32⟩

[75] Andrew S. Tanenbaum. *Modern Operating Systems*. Prentice Hall, Upper Saddle River, New Jersey, 1992. ⟨179⟩

[76] Donald E. Thomas and Philip R. Moorby. *The Verilog Hardware Description Language*. Kluwer, Boston, Massachusetts, fourth edition, 1998. ⟨32⟩

[77] Kenneth J. Turner, editor. *Using Formal Description Techniques: An Introduction to Estelle, Lotos, and SDL*. John Wiley & Sons, New York, 1993. ⟨240⟩

[78] The Unicode Consortium. *The Unicode Standard, Version 2.0*. Addison-Wesley, Reading, Massachusetts, 1996. ⟨166⟩

[79] Tiziano Villa, Timothy Kam, Robert K. Brayton, and Alberto Sangiovanni-Vincentelli. *Synthesis of Finite State Machines: Logic Optimization*. Kluwer, Boston, Massachusetts, 1997. ⟨26⟩

[80] John F. Wakerly. *Digital Design: Principles and Practices*. Prentice Hall, Upper Saddle River, New Jersey, second edition, 1994. ⟨19, 26⟩

[81] Neil H. E. Weste and Kamran Eshraghian. *Principles of CMOS VLSI Design: A Systems Perspective*. Addison-Wesley, Reading, Massachusetts, second edition, 1993. ⟨19⟩

[82] G. Winskel. *The Formal Semantics of Programming Languages: An Introduction*. Foundations of Computing. MIT Press, Cambridge, Massachusetts, 1993. ⟨3⟩

Index

Italic denote figures;
Bold denotes a definition.

This book was set by the author in Robert Slimbach's Minion and Charles Bigelow and Chris Holmes' Lucida Sans Typewriter using LaTeX2e, MetaPost, and Adobe Illustrator on a variety of computers running the Linux operating system.